Third Edition

Student Solutions Manual for
Technical Calculus
with Analytic Geometry

Peter Kuhfittig

Milwaukee School of Engineering

Brooks/Cole Publishing Company
Pacific Grove, California

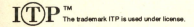

Brooks/Cole Publishing Company
A Division of Wadsworth, Inc.

Printed in the United States of America

10   9   8   7   6   5   4   3   2   1

ISBN 0-534-21853-9

Sponsoring Editor: Audra C. Silverie
Editorial Assistant: Carol Ann Benedict
Production Coordinator: Dorothy Bell
Cover Design: Lisa Berman
Cover Photo: Timothy Hursley
Printing and Binding: Malloy Lithographing, Inc.

Contents

CHAPTER 1   INTRODUCTION TO ANALYTIC GEOMETRY

Section 1.1

1. Let $(x_2,y_2) = (2,4)$ and $(x_1,y_1) = (5,2)$.   From the distance formula

$$d = \sqrt{(x_2 - x_1)^2 + (y_2 - y_1)^2}$$

we get

$$d = \sqrt{(2 - 5)^2 + (4 - 2)^2} = \sqrt{(-3)^2 + 2^2} = \sqrt{9 + 4} = \sqrt{13}$$

5. $d = \sqrt{(\sqrt{3} - 0)^2 + (4 - 2)^2} = \sqrt{3 + 4} = \sqrt{7}$

9. $d = \sqrt{[-10 - (-12)]^2 + (-2 - 0)^2} = \sqrt{2^2 + (-2)^2} = \sqrt{8}$

$$= \sqrt{2 \cdot 4} = 2\sqrt{2}$$

17. The points $(12,0)$, $(-4,8)$, and $(-1,-13)$ are all $5\sqrt{5}$ units from $(1,-2)$.

19. Distance from $(-1,-1)$ to $(2,8)$:

$$\sqrt{(-1 - 2)^2 + (-1 - 8)^2} = \sqrt{9 + 81} = \sqrt{90} = \sqrt{9 \cdot 10} = 3\sqrt{10}$$

Distance from $(2,8)$ to $(5,17)$:

$$\sqrt{(5 - 2)^2 + (17 - 8)^2} = \sqrt{90} = 3\sqrt{10}$$

Distance from $(-1,-1)$ to $(5,17)$:

$$\sqrt{18^2 + 6^2} = \sqrt{360} = 6\sqrt{10}$$

Total distance $6\sqrt{10} = 3\sqrt{10} + 3\sqrt{10}$, the sum of the other two distances.

21. Distance from (x,y) to y-axis: x units

Distance from (x,y) to $(2,0)$:

$$\sqrt{(x - 2)^2 + (y - 0)^2} = \sqrt{(x - 2)^2 + y^2}$$

By assumption,

$$\sqrt{(x - 2)^2 + y^2} = x$$
$$(x - 2)^2 + y^2 = x^2 \quad \text{(squaring both sides)}$$
$$x^2 - 4x + 4 + y^2 = x^2$$
$$y^2 - 4x + 4 = 0$$

1

1. Let $(x_2, y_2) = (1,7)$ and $(x_1, y_1) = (2,6)$. Then, by formula (1.3),

$$m = \frac{y_2 - y_1}{x_2 - x_1}$$

   we get

$$m = \frac{7 - 6}{1 - 2} = \frac{1}{-1} = -1$$

5. $m = \frac{8 - (-4)}{7 - (-3)} = \frac{8 + 4}{7 + 3} = \frac{12}{10} = \frac{6}{5}$

9. $m = \frac{-5 - 4}{3 - 3} = \frac{-9}{0}$ (undefined)

13. $m = \frac{-6 - 0}{0 - 8} = \frac{-6}{-8} = \frac{3}{4}$

17. Slope of given line $= \frac{1 - (-5)}{-7 - 6} = \frac{6}{-13} = -\frac{6}{13}$.

   Slope of perpendicular is therefore given by $-\frac{1}{(-6/13)} = \frac{13}{6}$.

21. Slope of line through $(0,-3)$ and $(-2,3)$: $\frac{-3 - 3}{0 - (-2)} = \frac{-6}{2} = -3$

   Slope of line through $(7,6)$ and $(9,0)$: $\frac{6 - 0}{7 - 9} = \frac{6}{-2} = -3$

   Slope of line through $(-2,3)$ and $(7,6)$: $\frac{3 - 6}{-2 - 7} = \frac{-3}{-9} = \frac{1}{3}$

   Slope of line through $(0,-3)$ and $(9,0)$: $\frac{-3 - 0}{0 - 9} = \frac{1}{3}$

   Since $-\frac{1}{-3} = \frac{1}{3}$, adjacent sides are perpendicular and opposite sides are parallel.

25. Midpoint: $\left(\frac{5 + 9}{2}, \frac{0 + 4}{2}\right) = (7,2)$

29. Midpoint: $\left(\frac{-3 + 9}{2}, \frac{-2 + 0}{2}\right) = (3,-1)$

   Slope of line through $(5,6)$ and $(3,-1)$: $\frac{6 - (-1)}{5 - 3} = \frac{7}{2}$

33. Slope of line through $(-1,-1)$ and $(3,-5)$: $\frac{-1 - (-5)}{-1 - 3} = \frac{4}{-4} = -1$

   Slope of line through $(x,2)$ and $(4,-6)$: $\frac{2 + 6}{x - 4} = \frac{8}{x - 4}$

   Since the two slopes must be equal, we have

$$\frac{8}{x - 4} = -1$$

$$8 = -x + 4 \quad \text{multiplying both sides by } x - 4$$

$$x = -4$$

1. Since $(x_1, y_1) = (-7, 2)$ and $m = \frac{1}{2}$, we get

$$y - 2 = \frac{1}{2}(x + 7)$$
$$2y - 4 = x + 7 \qquad \text{clearing fractions}$$
$$0 = x - 2y + 7 + 4$$
$$x - 2y + 11 = 0$$

5. $$y - 0 = -\frac{1}{3}(x - 0) \qquad (x_1, y_1) = (0,0); \quad m = -\frac{1}{3}$$
$$3y = -x$$
$$x + 3y = 0$$

9. $m = \dfrac{4 - (-6)}{-3 - 3} = \dfrac{10}{-6} = -\dfrac{5}{3}$

Let $(x_1, y_1) = (-3, 4)$. Then

$$y - 4 = -\frac{5}{3}(x + 3)$$
$$3y - 12 = -5x - 15 \qquad \text{multiplying by 3}$$
$$5x + 3y + 3 = 0$$

13. $m = \dfrac{3 - 4}{2 - (-6)} = \dfrac{-1}{8} = -\dfrac{1}{8}$

$$y - 3 = -\frac{1}{8}(x - 2) \qquad (x_1, y_1) = (2, 3)$$
$$8y - 24 = -x + 2 \qquad \text{multiplying by 8}$$
$$x + 8y - 26 = 0$$

17. $(x_1, y_1) = (0, -2), \quad m = -\frac{1}{3}$

$$y + 2 = -\frac{1}{3}(x - 0)$$
$$3y + 6 = -x$$
$$x + 3y + 6 = 0$$

21. Since $2x = 3y$, $y = \frac{2}{3}x$. From the form $y = mx + b$, $m = \frac{2}{3}$ and $b = 0$. The line passes through the origin and has slope $\frac{2}{3}$. (See drawing in answer section.)

25.
$$2x - 3y = 1 \qquad\qquad 4x - 6y + 3 = 0$$
$$-3y = -2x + 1 \qquad\qquad -6y = -4x - 3$$
$$y = \frac{2}{3}x - \frac{1}{3} \qquad\qquad y = \frac{4}{6}x + \frac{3}{6}$$
$$y = \frac{2}{3}x + \frac{1}{2}$$

From the form $y = mx + b$, $m = \frac{2}{3}$ in both cases, so that the lines are parallel.

3

29. $x + 3y = 5$ $y - 3x - 2 = 0$

  $3y = -x + 5$ $y = 3x + 2$

  $y = -\frac{1}{3}x + \frac{5}{3}$

The slopes are $-\frac{1}{3}$ and 3, respectively. Since the slopes are negative reciprocals, the lines are perpendicular.

33. To find the coordinates of the point of intersection, solve the equations simultaneously:

$2x - 4y = 1$
$\underline{3x + 4y = 4}$
$5x\qquad\ \ = 5\qquad$ (adding)

$\qquad x = 1$

From the second equation, $3(1) + 4y = 4$, and $y = \frac{1}{4}$. So the point of intersection is $(1,\frac{1}{4})$. From the equation $5x + 7y + 3 = 0$, we get

$7y = -5x - 3$

$y = -\frac{5}{7}x - \frac{3}{7}$ $\qquad\qquad$ slope $= -\frac{5}{7}$

Thus $(x_1, y_1) = (1, \frac{1}{4})$ and $m = -\frac{5}{7}$. The desired line is

$y - \frac{1}{4} = -\frac{5}{7}(x - 1)$

To clear fractions, we multiply both sides by 28:

$\qquad\qquad 28y - 7 = -20(x - 1)$

$\qquad\qquad 28y - 7 = -20x + 20$

$\qquad 20x + 28y - 27 = 0$

37. From $F = kx$, we get $3 = k \cdot \frac{1}{2}$. Thus $k = 6$ and $F = 6x$.

41. $R = aT + b$

$51 = a \cdot 100 + b$ $\qquad\qquad R = 51, \quad T = 100$
$\underline{54 = a \cdot 400 + b}$
$-3 = -300a$ $\qquad\qquad\qquad$ subtracting

$a = \frac{-3}{-300} = 0.01$

From the first equation, $51 = a \cdot 100 + b$, we get

$\quad 51 = (0.01)(100) + b$ $\qquad\qquad a = 0.01$

$\quad\ b = 50$

So the formula $R = aT + b$ becomes $R = 0.01T + 50$

Section 1.4

5. Intercepts. If $x = 0$, then $y = 0$, and if $y = 0$, then $x = 0$. So the only intercept is the origin.

Symmetry. If we replace x by $-x$, we get $y^2 = -x$, which does not reduce to the given equation. So there is no symmetry with respect to the y-axis.

4

If y is replaced by -y, we get $(-y)^2 = x$, which reduces to $y^2 = x$, the given equation. It follows that the graph is symmetric with respect to the x-axis.

To check for symmetry with respect to the origin, we replace x by -x and y by -y: $(-y)^2 = -x$. The resulting equation, $y^2 = -x$, does not reduce to the given equation.

Asymptotes. Since the equation is not in the form of a fraction with a variable in the denominator, there are no asymptotes.

Extent. Solving the equation for y in terms of x, we get

$$y = \pm\sqrt{x}$$

Note that to avoid imaginary values, x cannot be negative. It follows that the extent is $x \geq 0$.

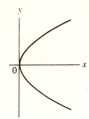

9. Intercepts. If either x = 0 or y = 0, we get 0 = 1, which is impossible. So there are no intercepts.

Symmetry. If x and y are replaced by -x and -y, respectively, we get $(-x)(-y) = 3$ or $xy = 3$. The graph is therefore symmetric with respect to the origin.

Asymptotes. Vertical: from $y = \frac{3}{x}$, we see that x = 0 (y-axis) is the vertical asymptote. Horizontal: if x gets large, then $\frac{3}{x}$ goes to 0. So y = 0 is the horizontal asymptote.

Extent. From $y = \frac{3}{x}$, we see that y is defined for all x except x = 0.

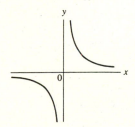

13. Intercepts. If x = 0, y = 0; if y = 0, then

$$x(x - 1)(x - 2)^2 = 0$$
$$x = 0, 1, 2$$

5

<u>Symmetry</u>. If x is replaced by -x, we get

$$y = -x(-x - 1)(-x - 2)^2,$$

which does not reduce to the given equation. So there is no symmetry with respect to the y-axis. Similarly, there is no other type of symmetry.

<u>Asymptotes</u>. None (the equation does not have the form of a fraction).

<u>Extent</u>. y is defined for all x.

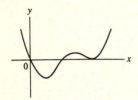

17. <u>Intercepts</u>. If x = 0, y = 0; if y = 0, we have

$$x\sqrt{1 - x^2} = 0$$

$$x = 0, \pm 1$$

<u>Symmetry</u>. If x and y are replaced by -x and -y, respectively, we get

$$-y = -x\sqrt{1 - (-x)^2},$$

which reduces to

$$y = x\sqrt{1 - x^2}.$$

The graph is therefore symmetric with respect to the origin.

<u>Asymptotes</u>. None (no fractions)

<u>Extent</u>. To avoid imaginary values, the radicand has to be greater than or equal to 0:

$$1 - x^2 \geq 0$$

$$1 \geq x^2 \qquad \text{adding } x^2 \text{ to both sides}$$

$$x^2 \leq 1$$

$$-1 \leq x \leq 1$$

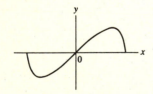

21. <u>Intercepts</u>. If $x = 0$, then
$$y = \frac{2}{(0 - 1)^2} = 2$$
If $y = 0$, then
$$0 = \frac{2}{(x - 1)^2} \qquad \text{(impossible)}$$
So the only intercept is $(0,2)$.

<u>Symmetry</u>. If x is replaced by $-x$, we get
$$y = \frac{2}{(-x - 1)^2}$$
which does not reduce to the given equation. Similarly, replacing y by $-y$ changes the equation. Consequently, there is no symmetry.

<u>Asymptotes</u>. Setting the denominator equal to 0, we get
$$(x - 1)^2 = 0 \qquad \text{or} \qquad x = 1$$
It follows that $x = 1$ is a vertical asymptote.

Also, as x gets large, y approaches 0. So the x-axis is a horizontal asymptote.

<u>Extent</u>. To avoid division by 0, x cannot be equal to 1. So the extent is all x except $x = 1$.

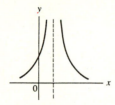

25. <u>Intercepts</u>. If $x = 0$, then
$$y = \frac{0 + 1}{(0 - 1)(0 + 2)} = \frac{1}{-2} = -\frac{1}{2}$$
If $y = 0$, we have
$$0 = \frac{x + 1}{(x - 1)(x + 2)}$$
which is possible only if $x + 1 = 0$, or $x = -1$.

<u>Symmetry</u>. If x is replaced by $-x$, we get
$$y = \frac{-x + 1}{(-x - 1)(-x + 2)}$$
which does not reduce to the given equation. In a similar manner, we can show that there is no symmetry with the x-axis or origin.

<u>Asymptotes</u>. Setting the denominator equal to 0, we get
$$(x - 1)(x + 2) = 0$$
$$x = 1, -2 \qquad \text{(vertical asymptotes)}$$
If x gets large

7

$$\frac{x + 1}{(x - 1)(x + 2)}$$

approaches 0 since the denominator is of higher degree than the numerator. It follows that the x-axis is a horizontal asymptote.

Extent. y is defined for all x except x = 1 and x = -2.

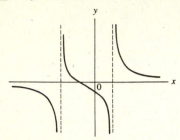

29. Intercepts. If x = 0, $y^2 = (-3)(5) = -15$, or $y = \pm\sqrt{15}\,j$, which is a pure imaginary number. If y = 0,

$$(x - 3)(x + 5) = 0$$
$$x = 3, -5$$

Symmetry. Replacing y by -y, we get $(-y)^2 = (x - 3)(x + 5)$, which reduces to the given equation. Hence the graph is symmetric with respect to the x-axis.

Asymptotes. None (no fractions).

Extent. From $y = \pm\sqrt{(x - 3)(x + 5)}$, we conclude that

$$(x - 3)(x + 5) \geq 0.$$

If $x \geq 3$, $(x - 3)(x + 5) \geq 0$. If $x \leq -5$, $(x - 3)(x + 5) \geq 0$, since both factors are negative (or zero). If $-5 < x < 3$, $(x - 3)(x + 5) < 0$. [For example, if x = 0, we get $(-3)(5) = -15$.] These observations are summarized in the following chart.

	test values	x - 3	x + 5	(x - 3)(x + 5)
x > 3	4	+	+	+
-5 < x < 3	0	-	+	-
x < -5	-6	-	-	+

Extent. $x \leq -5$, $x \geq 3$

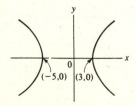

(-5,0) (3,0)

8

33. <u>Intercepts</u>. If $x = 0$, then $y^2 = \frac{-4}{-1} = 4$, or $y = \pm 2$. If $y = 0$, then

$$0 = \frac{x^2 - 4}{x^2 - 1}$$

which is possible only if $x^2 - 4 = 0$, or $x = \pm 2$.

<u>Symmetry</u>. The even powers on x and y tell us that if x is replaced by $-x$ and y is replaced by $-y$, the resulting equation will reduce to the given equation. The graph is therefore symmetric with respect to both axes.

<u>Asymptotes</u>. Vertical: setting the denominator equal to 0, we get

$$x^2 - 1 = 0 \qquad \text{or} \qquad x = \pm 1$$

Horizontal: dividing numerator and denominator by x^2, we get

$$y^2 = \frac{1 - \frac{4}{x^2}}{1 - \frac{1}{x^2}}$$

The right side approaches 1 as x gets large. Thus y^2 approaches 1, so that $y = \pm 1$ are the horizontal asymptotes.

<u>Extent</u>. From

$$y = \pm\sqrt{\frac{x^2 - 4}{x^2 - 1}}$$

we conclude that

$$\frac{x^2 - 4}{x^2 - 1} = \frac{(x - 2)(x + 2)}{(x - 1)(x + 1)} \geq 0$$

Since the signs change only at $x = 2$, -2, 1, and -1, we need to use arbitrary "test values" between. The results are summarized in the following chart:

	test values	x - 2	x - 1	x + 1	x + 2	$\frac{(x - 2)(x + 2)}{(x - 1)(x + 1)}$
x > 2	3	+	+	+	+	+
1 < x < 2	3/2	-	+	+	+	-
-1 < x < 1	0	-	-	+	+	+
-2 < x < -1	-3/2	-	-	-	+	-
x < -2	-3	-	-	-	-	+

Note that the fraction is positive only when $x > 2$, $-1 < x < 1$, and $x < -2$. Since $y = 0$ when $x = \pm 2$, the extent is $x \geq 2$, $-1 < x < 1$, $x \leq -2$.

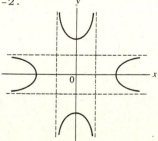

9

37. <u>Intercepts</u>. If t = 0, S = 0; if S = 0, we get
$$0 = 60t - 5t^2$$
$$0 = 5t(12 - t)$$
or t = 0, 12

<u>Symmetry</u>. None

<u>Asymptotes</u>. None

<u>Extent</u>. t ≥ 0 by assumption (graph is given in answer section).

<u>Section 1.5</u>

1. If y = 0, then
$$x^2(x - 1)(x - 2) = 0$$
Setting each factor equal to 0, we get
x = 0, 1, 2

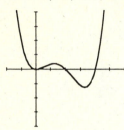

[−1, 3] by [−2, 2]

5. $x^4 - 2x^3 = 0$
$x^3(x - 2) = 0$
x = 0, 2

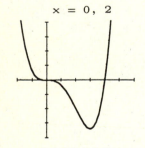

[−1, 3] by [−2, 2]

9. Domain: x ≥ 0 (to avoid imaginary values)

Vertical asymptotes: none (The denominator is always positive, that is, $1 + \sqrt{x} \neq 0$.)

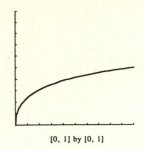

[0, 1] by [0, 1]

13. To find the vertical asymptotes, we set the denominator equal to 0:

$$2x^2 - 3 = 0$$
$$2x^2 = 3$$
$$x^2 = \frac{3}{2} \cdot \frac{2}{2} = \frac{6}{4}$$
$$x = \pm\frac{\sqrt{6}}{2}$$

Domain: y is defined for all x except $x = \pm\frac{\sqrt{6}}{2}$.

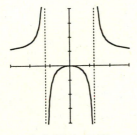

[-3, 3] by [-20, 20]

17. See graph in answer section.

21. See graph in answer section.

Section 1.7

1. Since (h,k) = (0,0) and r = 5, we get from the form
$$(x - h)^2 + (y - k)^2 = r^2$$
the equation
$$x^2 + y^2 = 25$$

3. $r^2 = (0 + 6)^2 + (0 - 8)^2 = 100$
$x^2 + y^2 = 100$

11

5.
$$(x + 2)^2 + (y - 5)^2 = 1^2$$
$$x^2 + 4x + 4 + y^2 - 10y + 25 = 1$$
$$x^2 + y^2 + 4x - 10y + 28 = 0$$

9. Diameter: distance from $(-2,-6)$ to $(1,5)$. Hence

$$r = \tfrac{1}{2}\sqrt{(-2 - 1)^2 + (-6 - 5)^2} = \tfrac{1}{2}\sqrt{9 + 121} = \tfrac{1}{2}\sqrt{130}$$
$$r^2 = \tfrac{1}{4}(130) = \tfrac{65}{2}$$

Center: midpoint of the line segment, whose coordinates are
$$\left(\frac{-2 + 1}{2}, \frac{-6 + 5}{2}\right) = \left(-\tfrac{1}{2}, -\tfrac{1}{2}\right)$$

Thus
$$(x + \tfrac{1}{2})^2 + (y + \tfrac{1}{2})^2 = \tfrac{65}{2}$$
$$x^2 + x + \tfrac{1}{4} + y^2 + y + \tfrac{1}{4} = \tfrac{65}{2}$$
$$x^2 + y^2 + x + y - 32 = 0$$

11.
$$x^2 + y^2 - 2x - 2y - 2 = 0$$
$$x^2 - 2x + y^2 - 2y = 2$$

We now add to each side the square of one-half the coefficient of x:

$$\left[\tfrac{1}{2}(-2)\right]^2 = 1$$
$$x^2 - 2x + \underline{1} + y^2 - 2y = 2 + \underline{1}$$

Similarly, we add 1 (the square of one-half the coefficient of y):
$$(x^2 - 2x + 1) + (y^2 - 2y + \underline{1}) = 2 + 1 + \underline{1}$$
$$(x - 1)^2 + (y - 1)^2 = 4$$

Center: $(h,k) = (1,1)$; radius: $\sqrt{4} = 2$

13.
$$x^2 + y^2 + 4x - 8y + 4 = 0$$
$$x^2 + 4x + y^2 - 8y = -4$$

Since
$$\left(\tfrac{1}{2} \cdot 4\right)^2 = 4 \quad \text{and} \quad \left[\tfrac{1}{2}(-8)\right]^2 = 16$$

we get
$$(x^2 + 4x + \underline{4}) + (y^2 - 8y + \underline{16}) = -4 + \underline{4} + \underline{16}$$
$$(x + 2)^2 + (y - 4)^2 = 16$$

The equation can be written
$$[x - (-2)]^2 + (y - 4)^2 = 4^2$$

It follows that
$$(h,k) = (-2,4) \quad \text{and} \quad r = 4$$

17.
$$x^2 + y^2 + 4x - 2y - 4 = 0$$
$$x^2 + 4x + y^2 - 2y = 4$$

Note that
$$\left(\tfrac{1}{2} \cdot 4\right)^2 = 4 \quad \text{and} \quad \left[\tfrac{1}{2}(-2)\right]^2 = 1$$
Adding 4 and 1, respectively, we get
$$(x^2 + 4x + 4) + (y^2 - 2y + 1) = 4 + 4 + 1$$
$$(x + 2)^2 + (y - 1)^2 = 9$$
The equation can be written
$$[x - (-2)]^2 + (y - 1)^2 = 3^2$$
So the center is $(-2,1)$ and the radius is 3.

21.
$$x^2 + y^2 - 4x + y + \tfrac{9}{4} = 0$$
$$x^2 - 4x \quad + y^2 + y \quad = -\tfrac{9}{4}$$

Note that
$$\left[\tfrac{1}{2}(-4)\right]^2 = 4 \quad \text{and} \quad \left(\tfrac{1}{2} \cdot 1\right)^2 = \tfrac{1}{4}$$

Adding 4 and $\tfrac{1}{4}$, respectively, we get
$$(x^2 - 4x + 4) + (y^2 + y + \tfrac{1}{4}) = -\tfrac{9}{4} + 4 + \tfrac{1}{4}$$
$$(x - 2)^2 + (y + \tfrac{1}{2})^2 = 2$$
The equation can be written
$$(x - 2)^2 + \left[y - \left(-\tfrac{1}{2}\right)\right]^2 = \left(\sqrt{2}\right)^2$$
Center: $\left(2, -\tfrac{1}{2}\right)$; radius: $\sqrt{2}$

25.
$$4x^2 + 4y^2 - 20x - 4y + 26 = 0$$
$$x^2 + y^2 - 5x - y + \tfrac{26}{4} = 0 \qquad \text{dividing by 4}$$
$$x^2 - 5x \quad + y^2 - y \quad = -\tfrac{26}{4}$$
$$x^2 - 5x + \tfrac{25}{4} + y^2 - y + \tfrac{1}{4} = -\tfrac{26}{4} + \tfrac{25}{4} + \tfrac{1}{4}$$
$$(x - \tfrac{5}{2})^2 + (y - \tfrac{1}{2})^2 = 0$$
Locus is the single point $\left(\tfrac{5}{2}, \tfrac{1}{2}\right)$

29.
$$x^2 + y^2 - 6x - 8y + 30 = 0$$
$$x^2 - 6x \quad + y^2 - 8y \quad = -30$$
$$x^2 - 6x + 9 + y^2 - 8y + 16 = -30 + 9 + 16$$
$$(x - 3)^2 + (y - 4)^2 = -5$$
(imaginary circle)

Section 1.8

1. Since the focus is on the x-axis, the form is $y^2 = 4px$. Since the focus is at $(3,0)$, $p = 3$ (positive). Thus $y^2 = 4(3)x$, or $y^2 = 12x$.

5. Since the focus is on the x-axis, the form is $y^2 = 4px$. The focus is on the left side of the origin, at $(-4,0)$. So $p = -4$ (negative). It follows that $y^2 = 4(-4)x$, or $y^2 = -16x$.

9. Since the directrix is $x = 2$, the focus is at $(-2,0)$. So the form is $y^2 = 4px$ with $p = -2$. Thus $y^2 = -8x$.

11. Since the focus is at $(0,-3)$, $p = -3$ (negative) and the form is $x^2 = 4py$. So $x^2 = 4(-3)y$, or $x^2 = -12y$.

13. Form: $y^2 = 4px$. Since the curve passes through $(-2,-4)$, the coordinates of this point must satisfy the equation. Substituting, we get

$$y^2 = 4px$$
$$(-4)^2 = 4p(-2) \qquad\qquad y = -4, \ x = -2$$
$$4p = -8$$
Thus $\quad\quad y^2 = -8x \qquad\qquad 4p = -8$

15. From $x^2 = 8y$, we have $x^2 = 4(2y)$. Thus $p = 2$ and the focus is at $(0,2)$.

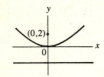

21. From $y^2 = -4x$, $y^2 = 4(-1x)$. So $p = -1$ and the focus is at $(-1,0)$.

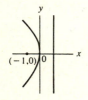

25. From $y^2 = 9x$,
$$y^2 = 4\left(\tfrac{9}{4}\right)x \qquad\qquad \text{Inserting } 4$$
So $p = \tfrac{9}{4}$ and the focus is at $\left(\tfrac{9}{4},0\right)$.

14

29.
$$3y^2 + 2x = 0$$
$$y^2 = -\frac{2}{3}x$$
$$y^2 = 4\left(-\frac{2}{3} \cdot \frac{1}{4}\right)x$$
$$y^2 = 4\left(-\frac{1}{6}\right)x$$

So the focus is at $\left(-\frac{1}{6}, 0\right)$.

33. We need to find the locus of points (x,y) equidistant from $(4,1)$ and the y-axis. Since the distance from (x,y) to the y-axis is x units, we get

$$\sqrt{(x - 4)^2 + (y - 1)^2} = x$$
$$(x - 4)^2 + (y - 1)^2 = x^2$$
$$x^2 - 8x + 16 + y^2 - 2y + 1 = x^2$$
$$y^2 - 2y - 8x + 17 = 0$$

37. If the origin is the lowest point on the cable, then the top of the right supporting tower is at $(100,70)$.

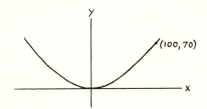

From the equation $x^2 = 4py$, we get
$$(100)^2 = 4p(70)$$
$$4p = \frac{10,000}{70} = \frac{1000}{7}$$
The equation is therefore
$$x^2 = \frac{1000}{7}y$$
To find the length of the cable 30 m from the center, we let $x = 30$:
$$30^2 = \frac{1000}{7}y \quad \text{and} \quad y = \frac{6300}{1000} = 6.3$$
So the length of the cable is $20 + 6.3 = 26.3$ m.

41. We place the vertex of the parabola at the origin, so that one point on the parabola is $(3,-3)$ (from the given dimensions). Substituting in the equation $x^2 = 4py$, we get
$$3^2 = 4p(-3)$$
$$4p = -3$$

15

The equation is therefore seen to be
$$x^2 = -3y$$

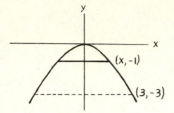

The right end of the beam 2 m above the base is at $(x,-1)$. To find x, let $y = -1$:
$$x^2 = -3(-1) = 3$$
$$x = \pm\sqrt{3}$$

Hence the length of the beam is $2|x| = 2\sqrt{3}$ m.

Section 1.9

1. The equation is
$$\frac{x^2}{25} + \frac{y^2}{16} = 1$$

So by (1.16), $a^2 = 25$ and $b^2 = 16$; thus $a = 5$ and $b = 4$. Since the major axis is horizontal, the vertices are at $(\pm5,0)$. From $b^2 = a^2 - c^2$,
$$16 = 25 - c^2$$
$$c^2 = 9$$
$$c = \pm3$$

The foci are therefore at $(\pm3,0)$, on the major axis. Finally, the length of the semiminor axis is equal to $b = 4$.

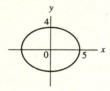

5. By (1.16), $a = 4$ and $b = 1$. From $b^2 = a^2 - c^2$,

$$1 = 16 - c^2$$
$$c^2 = 15$$
$$c = \pm\sqrt{15}$$

Since the major axis is horizontal, the vertices and foci lie on the x-axis. The vertices are therefore at $(\pm 4, 0)$ and the foci are at $(\pm\sqrt{15}, 0)$. The length of the semiminor axis is $b = 1$.

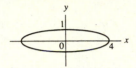

9. $$5x^2 + 2y^2 = 20$$
$$\frac{5x^2}{20} + \frac{2y^2}{20} = 1$$
$$\frac{x^2}{4} + \frac{y^2}{10} = 1$$

By (1.17), $a = \sqrt{10}$ and $b = 2$. Since the major axis is vertical, the vertices are at $(0, \pm\sqrt{10})$. From $b^2 = a^2 - c^2$

$$4 = 10 - c^2$$
$$c = \pm\sqrt{6}$$

The foci, also on the major axis, are therefore at $(0, \pm\sqrt{6})$, while the length of the semiminor axis is $b = 2$ (sketch in answer section).

13. $$x^2 + 2y^2 = 6$$
$$\frac{x^2}{6} + \frac{2y^2}{6} = 1$$
$$\frac{x^2}{6} + \frac{y^2}{3} = 1 \qquad \text{major axis horizontal}$$

Thus $a = \sqrt{6}$ and $b = \sqrt{3}$. From $b^2 = a^2 - c^2$,

$$3 = 6 - c^2$$
$$c = \pm\sqrt{3}$$

Vertices: $(\pm\sqrt{6}, 0)$; foci: $(\pm\sqrt{3}, 0)$. Length of semiminor axis: $b = \sqrt{3}$ (sketch in answer section).

17. $$2x^2 + 5y^2 = 50$$
$$\frac{2x^2}{50} + \frac{5y^2}{50} = 1$$
$$\frac{x^2}{25} + \frac{y^2}{10} = 1 \qquad \text{major axis horizontal}$$

Thus $a = 5$ and $b = \sqrt{10}$. From $b^2 = a^2 - c^2$,

$$10 = 25 - c^2$$
$$c^2 = 15$$

Vertices: $(\pm 5, 0)$; foci: $(\pm\sqrt{15}, 0)$; length of semiminor axis: $b = \sqrt{10}$.

21. Since the foci $(0, \pm 2)$ lie on the major axis, the major axis is vertical. So by (1.17) the form of the equation is

$$\frac{x^2}{b^2} + \frac{y^2}{a^2} = 1$$

Since the length of the major axis is 8, $a = 4$. From $b^2 = a^2 - c^2$ with $c = 2$,

$$b^2 = 16 - 4 = 12$$

Hence

$$\frac{x^2}{12} + \frac{y^2}{16} = 1 \qquad \text{or} \qquad 4x^2 + 3y^2 = 48$$

25. Since the vertices and foci are on the y-axis, the form of the equation is, by (1.17),

$$\frac{x^2}{b^2} + \frac{y^2}{a^2} = 1$$

From $b^2 = a^2 - c^2$ with $a = 8$ and $c = 5$,

$$b^2 = 64 - 25 = 39$$

Hence

$$\frac{x^2}{39} + \frac{y^2}{64} = 1$$

29. From the original derivation of the ellipse, $2a = 16$ and $a = 8$. Since the foci are at $(\pm 6, 0)$, $c = 6$. Thus

$$b^2 = a^2 - c^2 = 64 - 36 = 28$$

By (1.16) the equation is

$$\frac{x^2}{64} + \frac{y^2}{28} = 1$$

33. Distance from (x, y) to $(0, 0)$:
$$\sqrt{(x - 0)^2 + (y - 0)^2} = \sqrt{x^2 + y^2}$$

Distance from (x, y) to $(3, 0)$:
$$\sqrt{(x - 3)^2 + (y - 0)^2} = \sqrt{(x - 3)^2 + y^2}$$

From the given condition:
$$\sqrt{x^2 + y^2} = 2\sqrt{(x - 3)^2 + y^2}$$
$$x^2 + y^2 = 4[(x - 3)^2 + y^2] \qquad \text{squaring both sides}$$
$$x^2 + y^2 = 4(x^2 - 6x + 9 + y^2)$$
$$x^2 + y^2 = 4x^2 - 24x + 36 + 4y^2$$

$$0 = 3x^2 - 24x + 36 + 3y^2$$

$$3x^2 + 3y^2 - 24x + 36 = 0$$

$$x^2 + y^2 - 8x + 12 = 0$$

The locus is a circle.

37. Placing the center at the origin, the vertices are at $(\pm6,0)$. The road extends from $(-4,0)$ to $(4,0)$. Since the clearance is 4 m, the point $(4,4)$ lies on the ellipse, as shown.

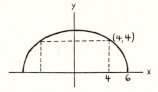

By (1.16),

$$\frac{x^2}{a^2} + \frac{y^2}{b^2} = 1$$

$$\frac{x^2}{36} + \frac{y^2}{b^2} = 1$$

To find b, we substitute the coordinates of $(4,4)$ in the equation:

$$\frac{16}{36} + \frac{16}{b^2} = 1$$

$$\frac{16}{b^2} = \frac{36}{36} - \frac{16}{36} = \frac{20}{36} = \frac{5}{9}$$

$$\frac{b^2}{16} = \frac{9}{5}$$

$$b^2 = \frac{(9)(16)}{5}$$

$$b = \frac{(3)(4)}{\sqrt{5}} = \frac{12}{\sqrt{5}} = \frac{12\sqrt{5}}{5}$$

So the height of the arch is $\frac{12\sqrt{5}}{5} = 5.4$ m to two significant digits.

Section 1.10

1. Comparing the given equation,

$$\frac{x^2}{16} - \frac{y^2}{9} = 1$$

to form (1.22), we see that the transverse axis is horizontal, with $a^2 = 16$ and $b^2 = 9$. So $a = 4$ and $b = 3$. From $b^2 = c^2 - a^2$, we get

$$9 = c^2 - 16$$

$$c = \pm5$$

19

So the vertices are at $(\pm 4,0)$ and the foci are at $(\pm 5,0)$. Using $a = 4$ and $b = 3$, we draw the auxiliary rectangle and sketch the curve:

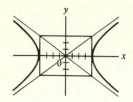

5. By (1.23), $a = 2$ and $b = 2$, transverse axis vertical. From $b^2 = c^2 - a^2$, $4 = c^2 - 4$ and $c = \pm\sqrt{8} = \pm 2\sqrt{2}$. So the vertices are at $(0,\pm 2)$ and the foci are at $(0,\pm 2\sqrt{2})$. Using $a = 2$ and $b = 2$, we draw the auxiliary rectangle and sketch the curve:

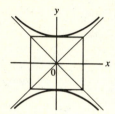

9. $2y^2 - 3x^2 = 24$

$$\frac{2y^2}{24} - \frac{3x^2}{24} = 1$$

$$\frac{y^2}{12} - \frac{x^2}{8} = 1$$

By (1.23), $a = \sqrt{12} = 2\sqrt{3}$ and $b = \sqrt{8} = 2\sqrt{2}$. From $b^2 = c^2 - a^2$, $8 = c^2 - 12$, so that $c = \pm\sqrt{20} = \pm 2\sqrt{5}$. Since the transverse axis lies along the y-axis, the vertices are at $(0,\pm 2\sqrt{3})$ and the foci at $(0,\pm 2\sqrt{5})$. Using $a = 2\sqrt{3}$ and $b = 2\sqrt{2}$, we draw the auxiliary rectangle:

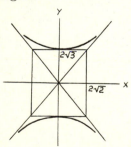

20

13. Since the foci (and hence the vertices) lie on the x-axis, the transverse axis is horizontal. By (1.22),

$$\frac{x^2}{a^2} - \frac{y^2}{b^2} = 1$$

Since the length of the transverse axis is 6, a = 3, and since the length of the conjugate axis is 4, b = 2. It follows that

$$\frac{x^2}{9} - \frac{y^2}{4} = 1 \qquad \text{or} \qquad 4x^2 - 9y^2 = 36$$

17. Since the foci lie in the y-axis, the transverse axis is vertical and by (1.23) the form is

$$\frac{y^2}{a^2} - \frac{x^2}{b^2} = 1$$

Since the length of the transverse axis is 12, a = 6. The foci are $(0, \pm 8)$, so that c = 8. To find b:

$$b^2 = c^2 - a^2 = 64 - 36 = 28$$

The equation is therefore

$$\frac{y^2}{36} - \frac{x^2}{28} = 1$$

21. By (1.22), the form of the equation is $\frac{x^2}{a^2} - \frac{y^2}{b^2} = 1$. Since a = 3 and c = 6, $b^2 = 36 - 9 = 27$. Thus

$$\frac{x^2}{9} - \frac{y^2}{27} = 1 \qquad \text{or} \qquad 3x^2 - y^2 = 27$$

25. By the original derivation of the equation of the hyperbola, 2a = 6 and a = 3. Since $(0, \pm 5)$ are the foci, c = 5. Thus $b^2 = 25 - 9 = 16$. By (1.23),

$$\frac{y^2}{a^2} - \frac{x^2}{b^2} = 1$$

$$\frac{y^2}{9} - \frac{x^2}{16} = 1 \qquad \text{or} \qquad 16y^2 - 9x^2 = 144$$

29. By (1.23), the equation has the form $\frac{y^2}{a^2} - \frac{x^2}{b^2} = 1$. Since a = 12, we have

$$\frac{y^2}{144} - \frac{x^2}{b^2} = 1$$

To find b, we substitute the coordinates of $(-1, 13)$ in the last equation:

$$\frac{169}{144} - \frac{1}{b^2} = 1$$

$$-\frac{1}{b^2} = \frac{144}{144} - \frac{169}{144} = -\frac{25}{144}$$

Thus $b^2 = \frac{144}{25}$. The equation is

$$\frac{y^2}{144} - \frac{x^2}{144/25} = 1$$

$$\frac{y^2}{144} - \frac{25x^2}{144} = 1$$

Section 1.11

5.
$$2x^2 - 3y^2 + 8x - 12y + 14 = 0$$
$$2x^2 + 8x - 3y^2 - 12y + 14 = 0$$
$$2(x^2 + 4x \quad) - 3(y^2 + 4y \quad) = -14 \qquad \text{factoring 2 and -3}$$

Note that the square of one-half the coefficient of x and y is
$$\left(\frac{1}{2} \cdot 4\right)^2 = 4$$
Inserting these values <u>inside</u> the parentheses and balancing the equation, we get
$$2(x^2 + 4x + \underline{4}) - 3(y^2 + 4y + \underline{4}) = -14 + 2 \cdot \underline{4} - 3 \cdot \underline{4}$$
$$2(x + 2)^2 - 3(y + 2)^2 = -18$$
$$\frac{3(y + 2)^2}{18} - \frac{2(x + 2)^2}{18} = 1$$
$$\frac{(y + 2)^2}{6} - \frac{(x + 2)^2}{9} = 1$$

The equation represents a hyperbola with transverse axis vertical.
Center: $(-2, -2)$, $a = \sqrt{6}$, $b = 3$.

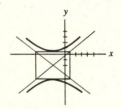

7.
$$16x^2 + 4y^2 + 64x - 12y + 57 = 0$$
$$16x^2 + 64x + 4y^2 - 12y + 57 = 0$$
$$16(x^2 + 4x \quad) + 4(y^2 - 3y \quad) = -57 \qquad \text{factoring 16 and 4}$$

Note that
$$\left(\frac{1}{2} \cdot 4\right)^2 = 4 \qquad \text{and} \qquad \left[\frac{1}{2}(-3)\right]^2 = \frac{9}{4}$$
Inserting these values inside the parentheses and balancing the equation, we get
$$16(x^2 + 4x + 4) + 4(y^2 - 3y + \frac{9}{4}) = -57 + 16 \cdot 4 + 4(\frac{9}{4})$$
$$16(x + 2)^2 + 4(y - \frac{3}{2})^2 = 16$$
$$\frac{(x + 2)^2}{1} + \frac{(y - 3/2)^2}{4} = 1$$

22

The equation represents an ellipse with major axis vertical.
Center: $(-2,\frac{3}{2})$, $a = 2$, $b = 1$.

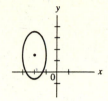

13.

$$64x^2 + 64y^2 - 16x - 96y - 27 = 0$$
$$64x^2 - 16x + 64y^2 - 96y - 27 = 0$$
$$64(x^2 - \tfrac{x}{4} \quad) + 64(y^2 - \tfrac{3y}{2} \quad) = 27$$
$$64(x^2 - \tfrac{x}{4} + \tfrac{1}{64}) + 64(y^2 - \tfrac{3y}{2} + \tfrac{9}{16}) = 27 + 1 + 36$$
$$64(x - \tfrac{1}{8})^2 + 64(y - \tfrac{3}{4})^2 = 64$$
$$(x - \tfrac{1}{8})^2 + (y \quad \tfrac{3}{4})^2 = 1$$

Circle of radius 1 centered at $(\frac{1}{8},\frac{3}{4})$.

17.

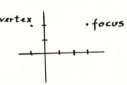

Distance from vertex to focus: $3 - (-1) = 4$. Thus $p = 4$. Since
the axis is horizontal, the form of the equation is
$(y - k)^2 = 4p(x - h)^2$. Thus

$$ (y - 2)^2 = 4\cdot4(x + 1) \qquad (h,k) = (-1,2), \ p = 4 $$
$$ (y - 2)^2 = 16(x + 1) $$

21.

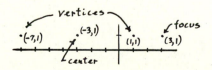

Distance between vertices is 8, so that $a = 4$. Center: $(-3,1)$
(point midway between vertices). Distance from center to one focus

23

is 6, so that c = 6. The transverse axis is horizontal, resulting in the form

$$\frac{(x - h)^2}{a^2} - \frac{(y - k)^2}{b^2} = 1$$

Since $b^2 = c^2 - a^2 = 36 - 16 = 20$, we get

$$\frac{(x + 3)^2}{16} - \frac{(y - 1)^2}{20} = 1 \qquad (h,k) = (-3,1)$$

23.

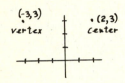

$(h,k) = (2,3)$

$a = 2 - (-3) = 5$ (distance from center to vertex)

$b = 2$ (length of minor axis is 4)

Form:

$$\frac{(x - h)^2}{a^2} + \frac{(y - k)^2}{b^2} = 1 \qquad \text{major axis horizontal}$$

Resulting equation:

$$\frac{(x - 2)^2}{25} + \frac{(y - 3)^2}{4} = 1 \qquad (h,k) = (2,3)$$

25. Form:

$$\frac{(x - h)^2}{a^2} - \frac{(y - k)^2}{b^2} = 1$$

Distance from center to vertex is 2, so that a = 2. From x - 2y = 1,

$$y = \tfrac{1}{2}x - \tfrac{1}{2}$$

So the slope m of one of the asymptotes is $\frac{1}{2}$. But $m = \frac{b}{a}$. Thus $\frac{1}{2} = \frac{b}{a} = \frac{b}{2}$ or b = 1. The equation is

$$\frac{(x - 1)^2}{4} - \frac{y^2}{1} = 1 \qquad (h,k) = (1,0)$$

29. Distance from vertex to focus: 4 - 1 = 3. Since the focus is to the left of the vertex, p = -3.

Form:

$$(y - k)^2 = 4p(x - h) \qquad \text{axis horizontal}$$

Equation:
$$(y + 2)^2 = -12(x - 4) \qquad (h,k) = (4,-2)$$

33.

$(-1,3) \cdot$ │ vertex

$(-1,1) \cdot$ │ center

$(-1,-2) \cdot$ │ focus

$(h,k) = (-1,1)$

$a = 3 - 1 = 2$ (distance from center to vertex)

$c = 1 - (-2) = 3$ (distance from center to focus)

$b^2 = c^2 - a^2 = 9 - 4 = 5$

Form:

$$\frac{(y - k)^2}{a^2} - \frac{(x - h)^2}{b^2} = 1 \qquad \text{transverse axis vertical}$$

Equation:

$$\frac{(y - 1)^2}{4} - \frac{(x + 1)^2}{5} = 1 \qquad (h,k) = (-1,1)$$

REVIEW EXERCISES FOR CHAPTER 1

1. Slope of line segment joining $(3,10)$ and $(7,4)$: $-\frac{3}{2}$

Slope of line segment joining $(4,2)$ and $(7,4)$: $\frac{2}{3}$.

Since the slopes are negative reciprocals, the line segments are perpendicular.

5. Slope of line segment joining $(-1,5)$ and $(3,9)$: 1.
Slope of line segment joining $(3,1)$ and $(7,5)$: 1.
Slope of line segment joining $(-1,5)$ and $(3,1)$: -1.
Slope of line segment joining $(3,9)$ and $(7,5)$: -1.
Since opposite sides are parallel, the figure is a parallelogram.
Moreover, since the line segment joining $(-1,5)$ and $(3,1)$ is perpendicular to the line segment joining $(-1,5)$ and $(3,9)$, the figure must be a rectangle. Finally:
Length of line segment joining $(-1,5)$ and $(3,1) = 4\sqrt{2}$.
Length of line segment joining $(-1,5)$ and $(3,9) = 4\sqrt{2}$.
Thus the figure is a square.

9. $r^2 = (1 - 0)^2 + (-2 - 0)^2 = 1 + 4 = 5$. We now get
$(x - 1)^2 + (y + 2)^2 = 5$ or $x^2 + y^2 - 2x + 4y = 0$

13. Ellipse, major axis vertical, a = 4, b = 3. From $b^2 = a^2 - c^2$,

$$9 = 16 - c^2$$
$$c = \pm\sqrt{7}$$

Vertices: $(0,\pm4)$; foci: $(0,\pm\sqrt{7})$ (sketch in answer section).

17. Parabola, axis horizontal. From $y^2 = -3x$, we have

$$y^2 = 4\left(-\frac{3}{4}\right)x \qquad \text{Inserting 4}$$

Thus, $p = -\frac{3}{4}$, placing the focus at $(-\frac{3}{4},0)$ (sketch in answer section).

21.
$$16x^2 - 64x + 9y^2 + 18y = 71$$
$$16(x^2 - 4x \quad) + 9(y^2 + 2y \quad) = 71 \qquad \text{Factoring 16 and 9}$$
$$16(x^2 - 4x + 4) + 9(y^2 + 2y + 1) = 71 + 16\cdot4 + 9\cdot1$$
$$16(x - 2)^2 + 9(y + 1)^2 = 144$$
$$\frac{(x - 2)^2}{9} + \frac{(y + 1)^2}{16} = 1 \qquad \text{Dividing by 144}$$

Ellipse, center at $(2,-1)$, major axis vertical, a = 4, b = 3.

25. Form:
$$(x - h)^2 = 4p(y - k) \qquad \text{axis vertical}$$
Distance from vertex $(1,3)$ to directrix $y = 0$ is 3, so that $p = 3$.
The equation is
$$(x - 1)^2 = 4(3)(y - 3)$$
or
$$(x - 1)^2 = 12(y - 3)$$

29.

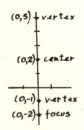

26

Center: $(0,2)$ (midway between vertices). Distance from center to vertex is 3, so that $a = 3$. Distance from center to focus is 4, so that $c = 4$.

$$b^2 = c^2 - a^2 = 16 - 9 = 7$$

Since the transverse axis is vertical, the form is

$$\frac{(y - k)^2}{a^2} - \frac{(x - h)^2}{b^2} = 1$$

and the equation is

$$\frac{(y - 2)^2}{9} - \frac{x^2}{7} = 1 \qquad (h,k) = (0,2)$$

33.

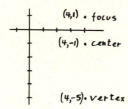

(4,1) . focus

(4,-1) . center

(4,-5) . vertex

$c = 2$ (distance from center to focus)

$a = 4$ (distance from center to vertex)

$b^2 = a^2 - c^2 = 16 - 4 = 12$

Form:

$$\frac{(x - h)^2}{b^2} + \frac{(y - k)^2}{a^2} = 1 \qquad \text{major axis vertical}$$

Equation:

$$\frac{(x - 4)^2}{12} + \frac{(y + 1)^2}{16} = 1 \qquad (h,k) = (4,-1)$$

37.   <u>Intercepts</u>. If $x = 0$, then $y = 0$; if $y = 0$, then $x(x - 4) = 0$

$$x = 0, 4$$

<u>Symmetry</u>. Replacing y by $-y$, we get

$$(-y)^2 = x(x - 4),$$

which reduces to the given equation. So the curve is symmetric with respect to the x-axis.

<u>Asymptotes</u>. None (equation is not in the form of a fraction).

<u>Extent</u>. Solving for y, we have

$$y = \pm\sqrt{x(x - 4)}$$

If $x > 4$, $x(x - 4) > 0$. If $0 < x < 4$, $x(x - 4) < 0$ [for example, if $x = 2$, we get $2(2 - 4) = -4$]. If $x < 0$, $x(x - 4) > 0$, since both factors are negative. So the extent is $x \leq 0$ and $x \geq 4$.

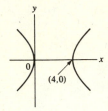

41. Placing the vertex at the origin, one point on the parabola is
     (0.90,0.60), as shown in the figure. The form is $y^2 = 4px$. To find
     p, we substitute the coordinates of the point in the equation:

$$(0.60)^2 = 4p(0.90)$$

$$p = \frac{(0.60)^2}{(4)(0.90)} = 0.10$$

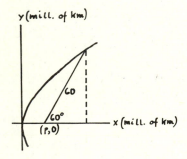

     To be at the focus, the light must be placed 0.10 ft from the vertex.

45. Placing the vertex at the origin, the problem is to find the focus
     (p,0).

     To find the point on the curve, consider the following triangle
     (lengths in millions of kilometers):

$$a = 60 \cos 60° = 60(\tfrac{1}{2}) = 30$$

$$b = 60 \sin 60° = 60(\tfrac{\sqrt{3}}{2}) = 30\sqrt{3}$$

So the point on the parabola has coordinates $(p+30, 30\sqrt{3})$. From $y^2 = 4px$ we get

$$(30\sqrt{3})^2 = 4p(p + 30)$$
$$2700 = 4p^2 + 120p$$
$$p^2 + 30p - 675 = 0$$
$$(p - 15)(p + 45) = 0$$
$$p = 15 \text{ (in millions of kilometers)}$$

Since the vertex is the point nearest the focus, the closest the comet comes to the sun is 15 million kilometers.

CHAPTER 2 INTRODUCTION TO CALCULUS: THE DERIVATIVE

<u>Section 2.1</u>

9. From the given equation $y = x + 2$, we see that y is defined for all
    x. From $x = y - 2$ it follows that x is defined for all y.

13. For the y-values to be real, the radicand must be positive:
    $$4 - x^2 \geq 0$$
    $$4 \geq x^2 \qquad \text{adding } x^2 \text{ to both sides}$$
    $$x^2 \leq 4$$
    $$-2 \leq x \leq 2 \qquad \text{domain}$$
    If $x = 0$, $y = \sqrt{4} = 2$; $y = 2$ is the largest possible value of y (if
    $x \neq 0$, $y < 2$). If $x = \pm 2$, then $y = 0$. So the range is $0 \leq y \leq 2$.

17. For $\sqrt{x - 2}$ to be real, we must have $x \geq 2$. If $x = 1$, then
    $y = (0)(j) = 0$. So the domain is the set of values $x = 1$ and
    $x \geq 2$.
    For the x-values under discussion, $y \geq 0$ (range).

21. $f(x) = 2x$
    $f(0) = 2 \cdot 0 = 0 \qquad x = 0$
    $f(6) = 2 \cdot 6 = 12 \qquad x = 6$

25. $f(x) = x^3 + 1$
    $f(0) = 0^3 + 1 = 1$; $f(1) = 1^3 + 1 = 2$; $f(-2) = (-2)^3 + 1 = -7$.

29. $G(z) = \sqrt{z^2 - 1}$. Leaving a blank space for x, we get
    $G(\ ) = \sqrt{(\ )^2 - 1}$
    Now fill in the blanks:
    $$G(a^2) = \sqrt{(a^2)^2 - 1} = \sqrt{a^4 - 1}$$
    $$G(x - 1) = \sqrt{(x - 1)^2 - 1} = \sqrt{x^2 - 2x}$$

33. $f(x) = x^2$; $g(x) = x + 1$
    $f(\ ) = (\ )^2$; $g(\ ) = (\ ) + 1$
    Filling in the blanks:
    (a) $f(g(x)) = (x + 1)^2 \qquad\qquad g(x) = x + 1$
    $\qquad\qquad\quad = x^2 + 2x + 1$

(b) $g(f(x)) = (x^2) + 1$ $f(x) = x^2$

              $= x^2 + 1$

(c) $f(f(x)) = (x^2)^2 = x^4$ $f(x) = x^2$

Section 2.2

1. Set the calculator in the radian mode and evaluate $\frac{\tan x}{x}$ for values near 0:

0.1	0.05	0.01	0.001
1.003	1.0008	1.00003	1.0000003

    The values approach 1.

5. Set the calculator in the radian mode and evaluate $\sec x - \tan x$ for values near $\frac{\pi}{2} = 1.570796$.

$1.5 \left(<\frac{\pi}{2}\right)$	$1.6 \left(>\frac{\pi}{2}\right)$	$1.57 \left(<\frac{\pi}{2}\right)$	$1.571 \left(>\frac{\pi}{2}\right)$	$1.5705 \left(<\frac{\pi}{2}\right)$
0.04	-0.01	4.0×10^{-4}	-1.02×10^{-4}	1.5×10^{-4}

    Based on these calculations, the limit appears to be 0.

9. The function $f(x) = x^4 - 3x + 2$ is defined at $x = 1$. As x approaches 1, $f(x)$ approaches $1^4 - 3(1) + 2 = 0$.

13. Since the function $f(x) = \frac{x^2 - x}{x}$ is undefined at $x = 0$, we evaluate the limit as follows:

    $$\lim_{x \to 0} \frac{x^2 - x}{x} = \lim_{x \to 0} \frac{x(x - 1)}{x} = \lim_{x \to 0} (x - 1) = -1$$

    (since x approaches 0 but is never equal to 0, division by 0 has been avoided).

17. The function is undefined at $x = 2$.

    $$\lim_{x \to 2} \frac{x^2 - 4x + 4}{x - 2} = \lim_{x \to 2} \frac{(x - 2)^2}{x - 2} = \lim_{x \to 2} (x - 2) = 0$$

21. The function is defined at $x = 4$. So we obtain the limit by inspection: as $x \to 4$

    $$\frac{x^2 + x - 8}{x + 4} \to \frac{4^2 + 4 - 8}{4 + 4} = \frac{12}{8} = \frac{3}{2}$$

25. $$\lim_{x \to 3} \frac{x^2 - 9}{3 - x} = \lim_{x \to 3} \frac{(x - 3)(x + 3)}{(-1)(x - 3)} = \lim_{x \to 3} \frac{x + 3}{-1} = -6$$

29. Since the function is defined at $x = 1$, we obtain the limit by inspection:

as $x \rightarrow 1$,

$$\frac{4x^3 - 2x}{x} \rightarrow 2$$

33. Dividing numerator and denominator by x^3, the highest power of x, we obtain

$$\lim_{x \to \infty} \frac{4 - 2/x^2 + 1/x^3}{5 + 3/x - 1/x^2} = \frac{4 - 0 + 0}{5 + 0 - 0} = \frac{4}{5}$$

37. Dividing numerator and denominator by x^2, the highest power of x, we obtain

$$\lim_{x \to \infty} \frac{1 - 4x}{x^2 + 1} = \lim_{x \to \infty} \frac{\frac{1}{x^2} - \frac{4}{x}}{1 + \frac{1}{x^2}} = \frac{0}{1} = 0$$

(The conclusion also follows from the fact that the denominator is of higher degree than the numerator.)

41. Divide numerator and denominator by x^2:

$$\lim_{x \to \infty} \frac{1/x^2 - 12}{2/x^2 + 6/x - 6} = \frac{-12}{-6} = 2$$

45. Rationalize the numerator:

$$\lim_{x \to \infty} \left(x - \sqrt{x^2 - 1} \right) = \lim_{x \to \infty} \frac{x - \sqrt{x^2 - 1}}{1} \cdot \frac{x + \sqrt{x^2 - 1}}{x + \sqrt{x^2 - 1}}$$

$$= \lim_{x \to \infty} \frac{x^2 - (\sqrt{x^2 - 1})^2}{x + \sqrt{x^2 - 1}} = \lim_{x \to \infty} \frac{x^2 - (x^2 - 1)}{x + \sqrt{x^2 - 1}}$$

$$= \lim_{x \to \infty} \frac{1}{x + \sqrt{x^2 - 1}} = 0$$

Section 2.4

1. To find $f(x + \Delta x)$, write

$$f(\quad) = 2(\quad) + 1$$

and fill in the blanks.

Step 1. $y + \Delta y = f(x + \Delta x) = 2(x + \Delta x) + 1$

Step 2. $\Delta y = 2(x + \Delta x) + 1 - (2x + 1)$
$$= 2x + 2\Delta x + 1 - 2x - 1 = 2\Delta x$$

Step 3. $\dfrac{\Delta y}{\Delta x} = \dfrac{2\Delta x}{\Delta x} = 2$

Step 4. $f'(x) = \lim\limits_{\Delta x \to 0} \dfrac{\Delta y}{\Delta x} = \lim\limits_{\Delta x \to 0} 2 = 2$

32

5. To find $f(x + \Delta x)$, write

$f(\quad) = (\quad)^2 + 1$

and fill in the blanks.

Step 1. $y + \Delta y = f(x + \Delta x) = (x + \Delta x)^2 + 1$

Step 2. $\Delta y = (x + \Delta x)^2 + 1 - (x^2 + 1)$
$= x^2 + 2x\Delta x + (\Delta x)^2 + 1 - x^2 - 1 = 2x\Delta x + (\Delta x)^2$

Step 3. $\dfrac{\Delta y}{\Delta x} = \dfrac{2x\Delta x + (\Delta x)^2}{\Delta x} = \dfrac{\Delta x(2x + \Delta x)}{\Delta x} = 2x + \Delta x$

Step 4. $f'(x) = \lim\limits_{\Delta x \to 0} \dfrac{\Delta y}{\Delta x} = \lim\limits_{\Delta x \to 0} (2x + \Delta x) = 2x$

9. Step 1. $y + \Delta y = f(x + \Delta x) = (x + \Delta x)^3 - 3(x + \Delta x)^2$

Step 2. $\Delta y = (x + \Delta x)^3 - 3(x + \Delta x)^2 - (x^3 - 3x^2)$
$= x^3 + 3x^2\Delta x + 3x(\Delta x)^2 + (\Delta x)^3 - 3x^2 - 6x\Delta x$
$\quad - 3(\Delta x)^2 - x^3 + 3x^2$
$= 3x^2\Delta x + 3x(\Delta x)^2 + (\Delta x)^3 - 6x\Delta x - 3(\Delta x)^2$
$= \Delta x[3x^2 + 3x\Delta x + (\Delta x)^2 - 6x - 3\Delta x]$

Step 3. $\dfrac{\Delta y}{\Delta x} = \dfrac{\Delta x[3x^2 + 3x\Delta x + (\Delta x)^2 - 6x - 3\Delta x]}{\Delta x}$

$= 3x^2 + 3x\Delta x + (\Delta x)^2 - 6x - 3\Delta x$

Step 4. $f'(x) = \lim\limits_{\Delta x \to 0} [3x^2 + 3x\Delta x + (\Delta x)^2 - 6x - 3\Delta x]$

$= 3x^2 - 6x$

13. Step 1. $y + \Delta y = f(x + \Delta x) = \dfrac{1}{(x + \Delta x)^2}$

Step 2. $\Delta y = \dfrac{1}{(x + \Delta x)^2} - \dfrac{1}{x^2}$

Note: The lowest common denominator is
$x^2(x + \Delta x)^2$
We therefore have

$\Delta y = \dfrac{x^2}{x^2(x + \Delta x)^2} - \dfrac{(x + \Delta x)^2}{x^2(x + \Delta x)^2}$

$= \dfrac{x^2 - (x + \Delta x)^2}{x^2(x + \Delta x)^2} = \dfrac{x^2 - x^2 - 2x\Delta x - (\Delta x)^2}{x^2(x + \Delta x)^2}$

$= \dfrac{\Delta x(-2x - \Delta x)}{x^2(x + \Delta x)^2}$

Step 3. $\dfrac{\Delta y}{\Delta x} = \dfrac{\Delta x(-2x - \Delta x)}{x^2(x + \Delta x)^2} \cdot \dfrac{1}{\Delta x} = \dfrac{-2x - \Delta x}{x^2(x + \Delta x)^2}$

Step 4. $f'(x) = \lim\limits_{\Delta x \to 0} \dfrac{-2x - \Delta x}{x^2(x + \Delta x)^2} = \dfrac{-2x}{x^2 x^2} = -\dfrac{2}{x^3}$

17. $\underline{\text{Step 1.}}$ $\quad y + \Delta y = f(x + \Delta x) = \sqrt{x + \Delta x}$

$\quad\underline{\text{Step 2.}}$ $\qquad \Delta y = \sqrt{x + \Delta x} - \sqrt{x}$

As in Example 4, we rationalize the numerator by multiplying both numerator and denominator by the quantity $\sqrt{x + \Delta x} - \sqrt{x}$:

$$\Delta y = \frac{\sqrt{x + \Delta x} - \sqrt{x}}{1} \cdot \frac{\sqrt{x + \Delta x} + \sqrt{x}}{\sqrt{x + \Delta x} + \sqrt{x}}$$

$$= \frac{(\sqrt{x + \Delta x})^2 - (\sqrt{x})^2}{\sqrt{x + \Delta x} + \sqrt{x}} = \frac{x + \Delta x - x}{\sqrt{x + \Delta x} + \sqrt{x}} = \frac{\Delta x}{\sqrt{x + \Delta x} + \sqrt{x}}$$

$\quad\underline{\text{Step 3.}}$ $\qquad \dfrac{\Delta y}{\Delta x} = \dfrac{\Delta x}{\sqrt{x + \Delta x} + \sqrt{x}} \cdot \dfrac{1}{\Delta x} = \dfrac{1}{\sqrt{x + \Delta x} + \sqrt{x}}$

$\quad\underline{\text{Step 4.}}$ $\quad f'(x) = \lim\limits_{\Delta x \to 0} \dfrac{1}{\sqrt{x + \Delta x} + \sqrt{x}} = \dfrac{1}{\sqrt{x} + \sqrt{x}} = \dfrac{1}{2\sqrt{x}}$

21. By Exercise 5, $f'(x) = 2x$. Thus $f'(-1) = 2(-1) = -2$; $f'(2) = 4$; $f'(3) = 6$.

Section 2.5

9. $\quad y = 5x^3 - 7x^2 + 2$

$\quad y' = 5(3x^2) - 7(2x) + 0$

$\qquad = 15x^2 - 14x$

13. $\quad y = \frac{1}{3}x^3 + \frac{1}{2}x^2 + x$

$\quad y' = \frac{1}{3}(3x^2) + \frac{1}{2}(2x) + 1$

$\qquad = x^2 + x + 1$

17. $\quad y = 20x^{10} - 24x^6 + 2x^3 - \sqrt{3}$

$\quad y' = 20(10x^9) - 24(6x^5) + 2(3x^2) + 0$

$\qquad = 200x^9 - 144x^5 + 6x^2$

(note that $\sqrt{3}$ is a constant)

21. $\quad y = \frac{1}{6}R^6 + \frac{1}{5}R^4 - \frac{1}{\sqrt{3}}$

$\quad \dfrac{dy}{dR} = \frac{1}{6}(6R^5) + \frac{1}{5}(4R^3) - 0$

$\qquad = R^5 + \frac{4}{5}R^3$

Section 2.6

1. $\quad s = 2t^2$ $\qquad\qquad\qquad\qquad v = 0$ when $4t = 0$ or $t = 0$

$\quad v = \dfrac{ds}{dt} = 2(2t) = 4t$

$\quad a = \dfrac{dv}{dt} = 4$

5. $s = 2t - t^2$ $v = 0$ when $2 - 2t = 0$

$$-2t = -2$$
$$t = 1$$

$$v = \frac{ds}{dt} = 2(1) - 2t = 2 - 2t$$

$$a = \frac{dv}{dt} = 0 - 2 = -2$$

9. $s = 3t^2 - 6t$ $v = 0$ when $6t - 6 = 0$

$$6t = 6$$
$$t = 1$$

$$v = \frac{ds}{dt} = 6t - 6$$

$$a = \frac{dv}{dt} = 6$$

13.   From $s = 50t^2$, we get $v = \dfrac{ds}{dt} = 50(2t)$

$$= 100t|_{t=10} = 1000 \ \frac{m}{s}$$

17.   The instantaneous rate of change of R with respect to T is $\dfrac{dR}{dT}$:

$$R = 20.0 + 0.520T + 0.00973T^2$$

$$\frac{dR}{dT} = 0 + 0.520(1) + 0.00973(2T)$$

$$= 0.520 + (0.00973)(2)T|_{T=125}$$

$$= 0.520 + (0.00973)(2)(125)$$

$$= 2.95 \ \frac{\Omega}{^\circ C}$$

21.   The instantaneous rate of change of S with respect to T is

$$\frac{dS}{dt} = 0 - 0.000085(2T)\Big|_{T=145}$$

$$= -0.000085(2)(145) = -0.025 \ lb/^\circ F$$

25. $i = \dfrac{dq}{dt} = \dfrac{d}{dt}(2t^2 + t) = 4t + 1$

29. $P = \dfrac{dW}{dt} = \dfrac{d}{dt}(5t^4 + 2t) = 20t^3 + 2|_{t=1} = 22 \ W$

Section 2.7

Group A

5. $y = x^5 - 3x^{-3} + 2x^{-2}$.   By (2.4),
      $y' = 5x^4 - 3(-3)x^{-4} + 2(-2)x^{-3} = 5x^4 + 9x^{-4} - 4x^{-3}$

9. $y = (2x^2 - 3)^4$.   By the generalized power rule
      $y' = 4(2x^2 - 3)^3 \dfrac{d}{dx}(2x^2 - 3) = 4(2x^2 - 3)^3(4x) = 16x(2x^2 - 3)^3$

13. $y = \dfrac{2}{\sqrt{2x^2 + 3}}$

We avoid the quotient rule by writing the function as follows:

$y = 2(2x^2 + 3)^{-1/2}$

$y' = 2(-\frac{1}{2})(2x^2 + 3)^{-3/2} \dfrac{d}{dx}(2x^2 + 3)$

by the generalized power rule. Thus

$y' = -(2x^2 + 3)^{-3/2}(4x)$

$\quad = -\dfrac{4x}{(2x^2 + 3)^{3/2}}$

17. $y = 2(x^3 - 3x)^{-1/2}$. By the generalized power rule,

$y' = 2(-\frac{1}{2})(x^3 - 3x)^{-3/2} \dfrac{d}{dx}(x^3 - 3x)$

$\quad = -(x^3 - 3x)^{-3/2}(3x^2 - 3)$

$\quad = -\dfrac{3x^2 - 3}{(x^3 - 3x)^{3/2}} = -\dfrac{3(x^2 - 1)}{(x^3 - 3x)^{3/2}}$

21. $y = x^3(x + 1)^2$

By the product rule:

$y' = x^3 \dfrac{d}{dx}(x + 1)^2 + (x + 1)^2 \dfrac{d}{dx}(x^3)$

$\quad = x^3 \cdot 2(x + 1) + (x + 1)^2(3x^2)$

Note that the terms on the right have common factor

$\quad\quad x^2(x + 1)$

so

$y' = x^2(x + 1)[x \cdot 2 + (x + 1)(3)]$

$\quad = x^2(x + 1)(5x + 3)$

25. $y = x^2(x^2 - 5)^2$

By the product rule,

$y' = x^2 \dfrac{d}{dx}(x^2 - 5)^2 + (x^2 - 5)^2 \dfrac{d}{dx}(x^2)$

Now by the generalized power rule,

$y' = x^2 \cdot 2(x^2 - 5)(2x) + (x^2 - 5)^2(2x)$

Factoring $2x(x^2 - 5)$, we get

$y' = 2x(x^2 - 5)[x^2 \cdot 2 + (x^2 - 5)]$

$\quad = 2x(x^2 - 5)(3x^2 - 5)$

29. $y = 2x^3(x - 2)^3$. By the product rule,

$y' = 2x^3 \dfrac{d}{dx}(x - 2)^3 + (x - 2)^3 \dfrac{d}{dx}(2x^3)$

$\quad = 2x^3 \cdot 3(x - 2)^2 + (x - 2)^3 \cdot 6x^2$

Factoring $6x^2(x - 2)^2$, we get

$$y' = 6x^2(x - 2)^2[x + (x - 2)] = 6x^2(x - 2)^2(2x - 2)$$
$$= 6x^2(x - 2)^2 \cdot 2(x - 1) = 12x^2(x - 2)^2(x - 1)$$

33. $P = \dfrac{t - 2}{t^2 + 4}$. By the quotient rule:

$$\frac{dP}{dt} = \frac{(t^2 + 4) \frac{d}{dt}(t - 2) - (t - 2) \frac{d}{dt}(t^2 + 4)}{(t^2 + 4)^2}$$

$$= \frac{(t^2 + 4)(1) - (t - 2)(2t)}{(t^2 + 4)^2}$$

$$= \frac{t^2 + 4 - 2t^2 + 4t}{(t^2 + 4)^2} = \frac{-t^2 + 4t + 4}{(t^2 + 4)^2}$$

37. $y = x(x^2 + 2)^{1/2}$. By the product rule:

$$y' = x \frac{d}{dx}(x^2 + 2)^{1/2} + (x^2 + 2)^{1/2} \frac{d}{dx}(x)$$

$$= x \cdot \frac{1}{2}(x^2 + 2)^{-1/2}(2x) + (x^2 + 2)^{1/2}(1)$$

$$= \frac{x^2}{(x^2 + 2)^{1/2}} + (x^2 + 2)^{1/2}$$

Since the common denominator is $(x^2 + 2)^{1/2}$, we need to write the second fraction as

$$\frac{(x^2 + 2)^{1/2}}{1} \cdot \frac{(x^2 + 2)^{1/2}}{(x^2 + 2)^{1/2}}$$

So

$$y' = \frac{x^2}{(x^2 + 2)^{1/2}} + \frac{(x^2 + 2)^{1/2}}{1} \frac{(x^2 + 2)^{1/2}}{(x^2 + 2)^{1/2}}$$

$$= \frac{x^2 + (x^2 + 2)^{1/2}(x^2 + 2)^{1/2}}{(x^2 + 2)^{1/2}} = \frac{x^2 + (x^2 + 2)}{(x^2 + 2)^{1/2}}$$

$$= \frac{2x^2 + 2}{(x^2 + 2)^{1/2}} = \frac{2(x^2 + 1)}{\sqrt{x^2 + 2}}$$

Group B

1. $y = 3x^{-2} - 2x^{-3}$
   $$y' = 3(-2x^{-3}) - 2(-3)x^{-4}$$
   $$= -6x^{-3} + 6x^{-4}$$

5. $y = \sqrt{1 - x} = (1 - x)^{1/2}$. By the power rule,

$$y' = \frac{1}{2}(1 - x)^{-1/2} \frac{d}{dx}(1 - x) = \frac{1}{2}(1 - x)^{-1/2}(-1) = -\frac{1}{2(1 - x)^{1/2}}$$

$$= - \frac{1}{2\sqrt{1 - x}}$$

9. $y = 4(x - x^2)^{-1/4}$. By the generalized power rule:

$$y' = 4(-\tfrac{1}{4})(x - x^2)^{-5/4} \frac{d}{dx}(x - x^2)$$

$$= -(x - x^2)^{-5/4}(1 - 2x)$$

$$= - \frac{1 - 2x}{(x - x^2)^{5/4}} = \frac{2x - 1}{(x - x^2)^{5/4}}$$

13. $y = \frac{x^2 - 3}{x - 2}$. By the quotient rule,

$$y' = \frac{(x - 2) \frac{d}{dx}(x^2 - 3) - (x^2 - 3) \frac{d}{dx}(x - 2)}{(x - 2)^2}$$

$$= \frac{(x - 2)(2x) - (x^2 - 3)(1)}{(x - 2)^2} = \frac{2x^2 - 4x - x^2 + 3}{(x - 2)^2} = \frac{x^2 - 4x + 3}{(x - 2)^2}$$

17. $y = x\sqrt{x^2 - 1} = x(x^2 - 1)^{1/2}$. By the product rule,

$$y' = x \frac{d}{dx}(x^2 - 1)^{1/2} + (x^2 - 1)^{1/2} \frac{d}{dx}(x)$$

$$= x\tfrac{1}{2}(x^2 - 1)^{-1/2}(2x) + (x^2 - 1)^{1/2} \qquad \text{generalized power rule}$$

$$= \frac{x^2}{\sqrt{x^2 - 1}} + \sqrt{x^2 - 1}$$

Since the common denominator is $\sqrt{x^2 - 1}$, we need to write the second term as

$$\frac{\sqrt{x^2 - 1}}{1} \cdot \frac{\sqrt{x^2 - 1}}{\sqrt{x^2 - 1}}$$

so

$$y' = \frac{x^2}{\sqrt{x^2 - 1}} + \frac{\sqrt{x^2 - 1}}{1} \cdot \frac{\sqrt{x^2 - 1}}{\sqrt{x^2 - 1}}$$

$$= \frac{x^2 + (x^2 - 1)}{\sqrt{x^2 - 1}} = \frac{2x^2 - 1}{\sqrt{x^2 - 1}}$$

21. $T = \theta^3(\theta + 7)^{1/2}$. By the product rule:

$$\frac{dT}{d\theta} = \theta^3 \frac{d}{d\theta}(\theta + 7)^{1/2} + (\theta + 7)^{1/2} \frac{d}{d\theta}(\theta^3)$$

$$= \theta^3 \cdot \tfrac{1}{2}(\theta + 7)^{-1/2} + (\theta + 7)^{1/2}(3\theta^2)$$

$$= \frac{\theta^3}{2(\theta + 7)^{1/2}} + 3\theta^2(\theta + 7)^{1/2}$$

Since the common denominator is $2(\theta + 7)^{1/2}$, we need to write the second fraction as

$$\frac{3\theta^2(\theta + 7)^{1/2}}{1} \cdot \frac{2(\theta + 7)^{1/2}}{2(\theta + 7)^{1/2}}$$

So

$$\frac{dT}{d\theta} = \frac{\theta^3}{2(\theta + 7)^{1/2}} + \frac{3\theta^2(\theta + 7)^{1/2}}{1} \frac{2(\theta + 7)^{1/2}}{2(\theta + 7)^{1/2}}$$

$$= \frac{\theta^3 + 3\theta^2(\theta + 7)^{1/2} \cdot 2(\theta + 7)^{1/2}}{2(\theta + 7)^{1/2}}$$

$$= \frac{\theta^3 + 6\theta^2(\theta + 7)}{2(\theta + 7)^{1/2}}$$

$$= \frac{\theta^3 + 6\theta^3 + 42\theta^2}{2(\theta + 7)^{1/2}} = \frac{7\theta^3 + 42\theta^2}{2(\theta + 7)^{1/2}}$$

$$= \frac{7\theta^2(\theta + 6)}{2\sqrt{\theta + 7}}$$

25. $y = \dfrac{x^2}{\sqrt{x + 1}} = \dfrac{x^2}{(x + 1)^{1/2}}$. By the quotient rule,

$$y' = \frac{(x + 1)^{1/2} \frac{d}{dx}(x^2) - x^2 \frac{d}{dx}(x + 1)^{1/2}}{[(x + 1)^{1/2}]^2}$$

$$= \frac{(x + 1)^{1/2}(2x) - x^2 \frac{1}{2}(x + 1)^{-1/2}}{x + 1} = \frac{2x\sqrt{x + 1} - \dfrac{x^2}{2\sqrt{x + 1}}}{x + 1}$$

To combine the expressions on top, note that the common denominator is $2\sqrt{x + 1}$:

$$= \frac{\dfrac{2x\sqrt{x + 1}}{1} \cdot \dfrac{2\sqrt{x + 1}}{2\sqrt{x + 1}} - \dfrac{x^2}{2\sqrt{x + 1}}}{x + 1} = \frac{\dfrac{4x(x + 1)}{2\sqrt{x + 1}} - \dfrac{x^2}{2\sqrt{x + 1}}}{x + 1}$$

$$= \frac{\dfrac{4x^2 + 4x - x^2}{2\sqrt{x + 1}}}{x + 1} = \frac{3x^2 + 4x}{2\sqrt{x + 1}} \cdot \frac{1}{x + 1} = \frac{3x^2 + 4x}{2(x + 1)\sqrt{x + 1}}$$

The expression

$$\frac{2x\sqrt{x + 1} - \dfrac{x^2}{2\sqrt{x + 1}}}{x + 1}$$

can also be simplified by clearing fractions: multiply numerator and denominator by $2\sqrt{x + 1}$ to reduce the complex fraction to an ordinary fraction. Thus

$$y' = \frac{2x\sqrt{x+1} - \dfrac{x^2}{2\sqrt{x+1}}}{x+1} \cdot \frac{2\sqrt{x+1}}{2\sqrt{x+1}}$$

$$= \frac{2x\sqrt{x+1} \cdot 2\sqrt{x+1} - x^2}{(x+1) \cdot 2\sqrt{x+1}}$$

$$= \frac{4x(x+1) - x^2}{2(x+1)\sqrt{x+1}} = \frac{3x^2 + 4x}{2(x+1)\sqrt{x+1}}$$

29. $y = \dfrac{x^2\sqrt{x}}{x^2+3} = \dfrac{x^2 x^{1/2}}{x^2+3} = \dfrac{x^{5/2}}{x^2+3}.$ By the quotient rule,

$$y' = \frac{(x^2+3)\left(\frac{5}{2}\right)x^{3/2} - x^{5/2}(2x)}{(x^2+3)^2}$$

$$= \frac{x^{3/2}\left[\left(\frac{5}{2}\right)(x^2+3) - x(2x)\right]}{(x^2+3)^2} \qquad \text{common factor } x^{3/2}$$

Now we multiply numerator and denominator by 2 (to clear fractions):

$$y' = \frac{x^{3/2}\left[\frac{5}{2}(x^2+3) - x(2x)\right]}{(x^2+3)^2} \cdot \frac{2}{2}$$

$$= \frac{x^{3/2}\left[5(x^2+3) - 4x^2\right]}{2(x^2+3)^2}$$

$$= \frac{x\sqrt{x}(5x^2 + 15 - 4x^2)}{2(x^2+3)^2} \qquad x^{3/2} = xx^{1/2} = x\sqrt{x}$$

$$= \frac{x\sqrt{x}(x^2+15)}{2(x^2+3)^2}$$

33. $y = \dfrac{x\sqrt{x-1}}{2x+3} = \dfrac{x(x-1)^{1/2}}{2x+3}.$ By the quotient rule,

$$y' = \frac{(2x+3)\frac{d}{dx}\left[x(x-1)^{1/2}\right] - x(x-1)^{1/2}\frac{d}{dx}(2x+3)}{(2x+3)^2}$$

$$= \frac{(2x+3)\left[x\left(\frac{1}{2}\right)(x-1)^{-1/2} + (x-1)^{1/2}\right] - x(x-1)^{1/2}(2)}{(2x+3)^2}$$

$$= \frac{(2x+3)\left[\dfrac{x}{2\sqrt{x-1}} + \sqrt{x-1}\right] - 2x\sqrt{x-1}}{(2x+3)^2} \cdot \frac{2\sqrt{x-1}}{2\sqrt{x-1}}$$

$$= \frac{(2x+3)\left[\dfrac{x}{2\sqrt{x-1}} + \sqrt{x-1}\right] \cdot 2\sqrt{x-1} - 2x\sqrt{x-1} \cdot 2\sqrt{x-1}}{(2x+3)^2 \cdot 2\sqrt{x-1}}$$

$$= \frac{(2x+3)[x + 2(x-1)] - 4x(\sqrt{x-1})^2}{2(2x+3)^2\sqrt{x-1}}$$

40

$$= \frac{(2x + 3)(3x - 2) - 4x(x - 1)}{2(2x + 3)^2\sqrt{x - 1}}$$

$$= \frac{6x^2 + 5x - 6 - 4x^2 + 4x}{2(2x + 3)^2\sqrt{x - 1}} = \frac{2x^2 + 9x - 6}{2(2x + 3)^2\sqrt{x - 1}}$$

37. By the product rule,

$$y' = (x^3 - 1)(2x + 3) + (3x^2)(x^2 + 3x + 2)\big|_{x=1}$$
$$= 0 + (3)(1 + 3 + 2) = 18$$

41. $\quad i = \dfrac{dq}{dt} = \dfrac{d}{dt}\dfrac{t}{t^2 + 4} = \dfrac{(t^2 + 4)(1) - t(2t)}{(t^2 + 4)^2}$

$$= \frac{t^2 + 4 - 2t^2}{(t^2 + 4)^2} = \frac{4 - t^2}{(t^2 + 4)^2} = 0$$

Hence

$$4 - t^2 = 0$$
$$t = \pm 2$$

Taking only the positive root, $t = 2$ s.

Section 2.8

1.

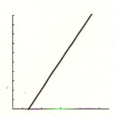

$[1.45, 1.55]$ by $[2.05, 2.15]$

5.

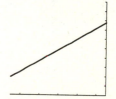

$[-2.05, -1.95]$ by $[0.88, 0.98]$

9.

$[3.95, 4.05]$ by $[3.95, 4.05]$

13.

$[2.45, 2.55]$ by $[1.51, 1.61]$

29.

[1.25,1.35] by [2.02,2.12]

Section 2.9

1. $2x + 3y = 3$

$2 + 3\frac{dy}{dx} = 0$

$\frac{dy}{dx} = -\frac{2}{3}$

5. $2x^2 - 3y^2 = 0$

$4x - 6y\frac{dy}{dx} = 0$

$-6y\frac{dy}{dx} = -4x$

$\frac{dy}{dx} = \frac{-4x}{-6y} = \frac{2x}{3y}$

9. $4y^4 - 3x^3 + 2 = 0$

$16y^3\frac{dy}{dx} - 9x^2 = 0$

$16y^3\frac{dy}{dx} = 9x^2$

$\frac{dy}{dx} = \frac{9x^2}{16y^3}$

13.

$$\frac{x^2}{a^2} + \frac{y^2}{b^2} = 1$$

$$\frac{1}{a^2}x^2 + \frac{1}{b^2}y^2 = 1$$

$$\frac{1}{a^2}(2x) + \frac{1}{b^2}(2y)\frac{dy}{dx} = 0$$

$$\frac{2y}{b^2}\frac{dy}{dx} = -\frac{2x}{a^2}$$

$$\frac{dy}{dx} = -\frac{2x}{a^2} \cdot \frac{b^2}{2y}$$

$$= -\frac{b^2x}{a^2y}$$

42

17.
$$x^2 y = 7$$

$$x^2 \frac{d}{dx}(y) + y \frac{d}{dx}(x^2) = 0 \qquad \text{product rule}$$

$$x^2 \frac{dy}{dx} + 2xy = 0$$

$$\frac{dy}{dx} = -\frac{2xy}{x^2} = -\frac{2y}{x}$$

21.
$$x^3 - 4x^2 y^2 + y^2 = 1$$

$$3x^2 - 4x^2 \frac{d}{dx}(y^2) + y^2 \frac{d}{dx}(-4x^2) + 2y \frac{dy}{dx} = 0 \qquad \text{product rule}$$

$$3x^2 - 4x^2(2y)\frac{dy}{dx} + y^2(-8x) + 2y\frac{dy}{dx} = 0$$

$$-8x^2 y \frac{dy}{dx} + 2y\frac{dy}{dx} = -3x^2 + 8xy^2$$

$$(-8x^2 y + 2y)\frac{dy}{dx} = 8xy^2 - 3x^2$$

$$\frac{dy}{dx} = \frac{8xy^2 - 3x^2}{2y - 8x^2 y}$$

25.
$$x^4 y^4 - 3y^2 + 5x = 6$$

$$x^4 \frac{d}{dx}(y^4) + y^4 \frac{d}{dx}(x^4) - 6y\frac{dy}{dx} + 5 = 0 \qquad \text{product rule}$$

$$x^4(4y^3)\frac{dy}{dx} + y^4(4x^3) - 6y\frac{dy}{dx} + 5 = 0$$

$$4x^4 y^3 \frac{dy}{dx} - 6y\frac{dy}{dx} = -4x^3 y^4 - 5$$

$$(4x^4 y^3 - 6y)\frac{dy}{dx} = -(4x^3 y^4 + 5)$$

$$\frac{dy}{dx} = -\frac{4x^3 y^4 + 5}{4x^4 y^3 - 6y}$$

29. Since $y = u^{p/q}$, we have $y^q = [u^{p/q}]^q = u^p$. From $y^q = u^p$, we get

$$qy^{q-1}\frac{dy}{dx} = pu^{p-1}\frac{du}{dx}$$

$$\frac{dy}{dx} = \frac{pu^{p-1}}{qy^{q-1}} \cdot \frac{du}{dx}$$

$$= \frac{pu^{p-1}}{q[u^{p/q}]^{q-1}} \cdot \frac{du}{dx} \qquad y = u^{p/q}$$

$$= \frac{pu^{p-1}}{qu^{p-p/q}} \cdot \frac{du}{dx} \qquad \text{multiplying exponents}$$

To subtract exponents, note that

$$(p - 1) - (p - p/q) = p - 1 - p + p/q = p/q - 1$$

Thus

$$\frac{dy}{dx} = \frac{p}{q} u^{p/q-1} \frac{du}{dx}$$

1. $y = 5x^4 + 5x^3 - 3x + 1$

    $y' = 20x^3 + 15x^2 - 3$

    $y'' = 60x^2 + 30x$

7. $f(x) = (5 + x)^{1/2}$

    $f'(x) = \frac{1}{2}(5 + x)^{-1/2}$

    $f''(x) = \left(\frac{1}{2}\right)\left(-\frac{1}{2}\right)(5 + x)^{-3/2}$

    $f'''(x) = \left(\frac{1}{2}\right)\left(-\frac{1}{2}\right)\left(-\frac{3}{2}\right)(5 + x)^{-5/2} = \dfrac{3}{8(5 + x)^{5/2}}$

9. $y = \dfrac{3 + 2x}{3 - 2x}$

    $\dfrac{dy}{dx} = \dfrac{(3 - 2x)(2) - (3 + 2x)(-2)}{(3 - 2x)^2} = \dfrac{6 - 4x + 6 + 4x}{(3 - 2x)^2}$

    $\quad = \dfrac{12}{(3 - 2x)^2} = 12(3 - 2x)^{-2}$

    $\dfrac{d^2y}{dx^2} = 12(-2)(3 - 2x)^{-3}(-2) = \dfrac{48}{(3 - 2x)^3}$

REVIEW EXERCISES FOR CHAPTER 2

1. $f(x) = x^2 - 1$

    $f(0) = 0^2 - 1 = -1 \qquad\qquad x = 0$

    $f(1) = 1^2 - 1 = 0 \qquad\qquad x = 1$

    $f(\sqrt{2}) = (\sqrt{2})^2 - 1 = 1 \qquad x = \sqrt{2}$

5. (a) To avoid imaginary values, we must have $x \geq 1$. The range is $y \geq 0$.

    (b) The cube root of any real number is a real number. So y is defined for all x.

9. $\displaystyle\lim_{x \to 0} \dfrac{x^3 - x^2 + 3x}{x} = \lim_{x \to 0} \dfrac{x(x^2 - x + 3)}{x} \qquad$ factoring x

    $\qquad\qquad = \displaystyle\lim_{x \to 0} (x^2 - x + 3) = 3$

13. Since the function is not defined at $x = 1$, we find the limit by rationalizing the numerator:

    $\displaystyle\lim_{x \to 1} \dfrac{\sqrt{x} - 1}{x - 1} = \lim_{x \to 1} \dfrac{\sqrt{x} - 1}{x - 1} \cdot \dfrac{\sqrt{x} + 1}{\sqrt{x} + 1} = \lim_{x \to 1} \dfrac{x - 1}{(x - 1)(\sqrt{x} + 1)}$

    $\qquad\qquad = \displaystyle\lim_{x \to 1} \dfrac{1}{\sqrt{x} + 1} = \dfrac{1}{2}$

17. By inspection, $\sqrt{x - 4} \to 0$ as $x \to 4$ from the right. (The restriction 4+ is necessary to avoid imaginary values.)

21. (a) <u>Step 1.</u> $y + \Delta y = f(x + \Delta x) = \dfrac{1}{4 - (x + \Delta x)}$

   <u>Step 2.</u> $\Delta y = \dfrac{1}{4 - (x + \Delta x)} - \dfrac{1}{4 - x}$

   Note that the common denominator is

   $$[4 - (x + \Delta x)](4 - x)$$

   We now get

   $$\Delta y = \frac{4 - x}{[4 - (x + \Delta x)](4 - x)} - \frac{4 - (x + \Delta x)}{[4 - (x + \Delta x)](4 - x)}$$

   $$= \frac{4 - x - 4 + (x + \Delta x)}{[4 - (x + \Delta x)](4 - x)} = \frac{\Delta x}{[4 - (x + \Delta x)](4 - x)}$$

   <u>Step 3.</u> $\dfrac{\Delta y}{\Delta x} = \dfrac{\Delta x}{[4 - (x + \Delta x)](4 - x)} \cdot \dfrac{1}{\Delta x}$

   $$= \frac{1}{[4 - (x + \Delta x)](4 - x)}$$

   <u>Step 4.</u> $f'(x) = \lim\limits_{\Delta x \to 0} \dfrac{1}{[4 - (x + \Delta x)](4 - x)}$

   $$= \frac{1}{(4 - x)^2}$$

(b) <u>Step 1.</u> $y + \Delta y = f(x + \Delta x) = \sqrt{x + \Delta x}$

   <u>Step 2.</u> $\Delta y = \sqrt{x + \Delta x} - \sqrt{x}$

   Rationalizing the numerator,

   $$\Delta y = \frac{\sqrt{x + \Delta x} - \sqrt{x}}{1} \cdot \frac{\sqrt{x + \Delta x} + \sqrt{x}}{\sqrt{x + \Delta x} + \sqrt{x}}$$

   $$= \frac{(\sqrt{x + \Delta x})^2 - (\sqrt{x})^2}{\sqrt{x + \Delta x} + \sqrt{x}} = \frac{x + \Delta x - x}{\sqrt{x + \Delta x} + \sqrt{x}} = \frac{\Delta x}{\sqrt{x + \Delta x} + \sqrt{x}}$$

   <u>Step 3.</u> $\dfrac{\Delta y}{\Delta x} = \dfrac{\Delta x}{\sqrt{x + \Delta x} + \sqrt{x}} \cdot \dfrac{1}{\Delta x} = \dfrac{1}{\sqrt{x + \Delta x} + \sqrt{x}}$

   <u>Step 4.</u> $f'(x) = \lim\limits_{\Delta x \to 0} \dfrac{1}{\sqrt{x + \Delta x} + \sqrt{x}} = \dfrac{1}{\sqrt{x} + \sqrt{x}} = \dfrac{1}{2\sqrt{x}}$

25. $y = \dfrac{x - 4}{x + 1}$. By the quotient rule,

   $$y' = \frac{(x + 1) - (x - 4)}{(x + 1)^2} = \frac{5}{(x + 1)^2}$$

29. $y = x\sqrt{4 - x^2} = x(4 - x^2)^{1/2}$. By the product rule,

   $$y' = x\frac{d}{dx}(4 - x^2)^{1/2} + (4 - x^2)^{1/2}\frac{d}{dx}(x)$$

   $$= x \cdot \frac{1}{2}(4 - x^2)^{-1/2}(-2x) + (4 - x^2)^{1/2}$$

   $$= \frac{-x^2}{(4 - x^2)^{1/2}} + (4 - x^2)^{1/2}$$

   $$= \frac{-x^2}{(4 - x^2)^{1/2}} + \frac{(4 - x^2)^{1/2}}{1} \cdot \frac{(4 - x^2)^{1/2}}{(4 - x^2)^{1/2}}$$

$$= \frac{-x^2 + (4 - x^2)}{(4 - x^2)^{1/2}}$$

$$= \frac{4 - 2x^2}{\sqrt{4 - x^2}}$$

33.
$$x^3 y + x y^3 = 5$$

$$x^3 \frac{dy}{dx} + y \frac{d}{dx}(x^3) + x \frac{d}{dx}(y^3) + y^3 \frac{d}{dx}(x) = 0 \qquad \text{product rule}$$

$$x^3 \frac{dy}{dx} + y(3x^2) + x(3y^2)\frac{dy}{dx} + y^3 = 0$$

$$(x^3 + 3xy^2)\frac{dy}{dx} = -(3x^2 y + y^3)$$

$$\frac{dy}{dx} = - \frac{3x^2 y + y^3}{x^3 + 3xy^2}$$

37. (a)
$$f(x) = \frac{x}{\sqrt{x - 1}} = \frac{x}{(x - 1)^{1/2}}$$

$$f'(x) = \frac{(x - 1)^{1/2} - x\left(\frac{1}{2}\right)(x - 1)^{-1/2}}{[(x - 1)^{1/2}]^2}$$

$$= \frac{\sqrt{x - 1} - \frac{x}{2\sqrt{x - 1}}}{x - 1} \cdot \frac{2\sqrt{x - 1}}{2\sqrt{x - 1}}$$

$$= \frac{2(x - 1) - x}{2(x - 1)\sqrt{x - 1}} = \frac{x - 2}{2(x - 1)\sqrt{x - 1}} = 0$$

Thus $f'(x) = 0$ when $x = 2$.

(b)
$$f(x) = 2x^3 - 6x^2 + 4$$

$$f'(x) = 6x^2 - 12x$$

$$f''(x) = 12x - 12 = 0 \text{ when } x = 1$$

41. $V = IR = (4.12 + 0.020t)(0.010t^2)$. By the product rule,
$$\frac{dV}{dt} = (4.12 + 0.020t)(0.020t) + (0.020)(0.010t^2)$$

Letting $t = 2.5\,s$, we get
$$\frac{dV}{dt} = 0.21 \frac{V}{s}$$

45. $F = \frac{1}{r^2} = r^{-2}$. Thus

$$\frac{dF}{dr} = -2r^{-3} = - \frac{2}{r^3}\bigg|_{r=5.02} = - \frac{2}{(5.02)^3} = -0.0158 \text{ dyne/cm}$$

(The negative sign indicates that F decreases as r increases.)

CHAPTER 3   APPLICATIONS OF THE DERIVATIVE

<u>Section 3.1</u>

1. Slope of tangent line: $y' = 2$

    Slope of normal line: $m = -\frac{1}{2}$

    Since the line passes through $(1,2)$, we get

    $$y - 2 = -\frac{1}{2}(x - 1)$$
    $$2y - 4 = -x + 1 \qquad \text{multiplying by 2}$$
    $$x + 2y - 5 = 0$$

    Note that the tangent line is $y = 2x$, same as the given line.

5. Slope of tangent line: $y' = 2x|_{x=1} = 2$. Since $m = 2$ and
    $(x_1, y_1) = (1,1)$, we get from the point-slope form:

    $$y - 1 = 2(x - 1)$$
    $$y - 1 = 2x - 2$$
    $$2x - y - 1 = 0$$

    For the normal line, $m = -\frac{1}{2}$. Thus

    $$y - 1 = -\frac{1}{2}(x - 1)$$
    $$2y - 2 = -x + 1 \qquad \text{multiplying by 2}$$
    $$x + 2y - 3 = 0$$

9. $y = \frac{1}{x} = x^{-1}$; $y' = -1x^{-2} = -\frac{1}{x^2}\Big|_{x=4} = -\frac{1}{16}$

    Tangent line:
    $$y - \frac{1}{4} = -\frac{1}{16}(x - 4)$$
    $$16y - 4 = -x + 4$$
    $$x + 16y - 8 = 0$$

    Normal line:
    $$y - \frac{1}{4} = 16(x - 4) \qquad m = -\frac{1}{(-1/16)}$$
    $$4y - 1 = 64x - 256$$
    $$64x - 4y - 255 = 0$$

13.
    $$x^2 + y^2 = 25$$

    $$2x + 2y\frac{dy}{dx} = 0 \qquad \text{differentiating implicitly}$$

    $$\frac{dy}{dx} = -\frac{x}{y}\Big|_{(-3,-4)} = -\frac{-3}{-4} = -\frac{3}{4}$$

Tangent Line:
$$y + 4 = -\frac{3}{4}(x + 3)$$
$$4y + 16 = -3x - 9$$
$$3x + 4y + 25 = 0$$

17.
$$3x^2 - xy + y^2 = 3$$
$$6x + \left(-x\frac{dy}{dx} - y\right) + 2y\frac{dy}{dx} = 0$$
$$-x\frac{dy}{dx} + 2y\frac{dy}{dx} = -6x + y$$
$$(-x + 2y)\frac{dy}{dx} = -6x + y$$
$$\frac{dy}{dx} = \frac{-6x + y}{-x + 2y}\Big|_{(1,1)} = \frac{-6 + 1}{-1 + 2} = -5$$

Tangent line:
$$y - 1 = -5(x - 1)$$
$$y - 1 = -5x + 5$$
$$5x + y - 6 = 0$$

Normal line:
$$y - 1 = \tfrac{1}{5}(x - 1) \qquad m = -\frac{1}{-5}$$
$$5y - 5 = x - 1$$
$$x - 5y + 4 = 0$$

Section 3.2

1. $y = x^2 - 2x + 1$ $y' = 2x - 2$
$$2x - 2 = 0$$
$$x = 1 \quad \text{(critical value)}$$

Substituting $x = 1$ in $y = x^2 - 2x + 1$, we get $y = 0$. So $(1,0)$ is
the critical point.
If $x < 1$, $y' < 0$; if $x > 1$, $y' > 0$. Since the function is decreasing
to the left of $x = 1$ and increasing to the right, $(1,0)$ is a minimum
point.

5. $y = 2x^3 + 3x^2 - 12x + 6$
$$y' = 6x^2 + 6x - 12$$
$$6x^2 + 6x - 12 = 0$$
$$6(x^2 + x - 2) = 0$$
$$6(x + 2)(x - 1) = 0 \qquad x = -2, 1$$

If x = -2, y = 26; if x = 1, y = -1. So (-2,26) and (1,-1) are the critical points.

To see where y′ is positive or negative, we substitute certain convenient values between the critical values.

	test values	x + 2	x - 1	$y' = 6(x + 2)(x - 1)$
x < -2	-3	-	-	+
-2 < x < 1	0	+	-	-
x > 1	2	+	+	+

Summary:

If x < -2, y′ > 0, so that f(x) is increasing

If -2 < x < 1, y′ < 0, so that f(x) is decreasing

If x > 1, y′ > 0, so that f(x) is increasing

We conclude that (-2,26) is a maximum point and (1,-1) is a minimum point.

9. $y = x^4 - 2x^2 - 2$

$$y' = 4x^3 - 4x$$
$$4x^3 - 4x = 0$$
$$4x(x^2 - 1) = 0$$
$$4x(x + 1)(x - 1) = 0$$
$$x = -1, 0, 1 \qquad \text{critical values}$$

To see where y′ < 0 and where y′ > 0, we substitute certain test values:

	test values	4x	x + 1	x - 1	$y' = 4x(x + 1)(x - 1)$
x < -1	-2	-	-	-	-
-1 < x < 0	-1/2	-	+	-	+
0 < x < 1	1/2	+	+	-	-
x > 1	2	+	+	+	+

Summary:

If $x < -1$, $y' < 0$: $f(x)$ is decreasing

If $-1 < x < 0$, $y' > 0$: $f(x)$ is increasing

If $0 < x < 1$, $y' < 0$: $f(x)$ is decreasing

If $x > 1$, $y' > 0$: $f(x)$ is increasing

It follows that $(\pm 1, -3)$ are minimum points and $(0, -2)$ is a maximum point.

13. $y = 4 - 4x^3 - 3x^4$

$$y' = -12x^2 - 12x^3$$
$$-12x^2 - 12x^3 = 0$$
$$-12x^2(1 + x) = 0$$
$$x = -1, \ 0$$

Critical points: $(-1,5)$ and $(0,4)$.

	test values	$-12x^2$	$1 + x$	$y' = -12x^2(1 + x)$
$x < -1$	-2	$-$	$-$	$+$
$-1 < x < 0$	$-1/2$	$-$	$+$	$-$
$x > 0$	1	$-$	$+$	$-$

Observe that to the left of $x = -1$, the function is increasing, and to the right of $x = -1$, the function is decreasing. It follows that $(-1,5)$ is a maximum.

Since the function is decreasing to the right of $x = -1$, it is decreasing to the left and right of $x = 0$. So the point $(0,4)$ is neither a minimum nor a maximum.

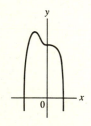

17. $s = 80t - 5t^2$

$\frac{ds}{dt} = 80 - 10t = 0$

$t = 8$

If $t < 8$, $\frac{ds}{dt} > 0$; if $t > 8$, $\frac{ds}{dt} < 0$. So s attains a maximum value at $t = 8$. The maximum altitude is

$s = 80t - 5t^2\big|_{t=8} = 320$ m

Section 3.3

1. $f(x) = 2x^2 - 4x$; $f'(x) = 4x - 4$; $f''(x) = 4$

Step 1. Critical points:

$f'(x) = 4x - 4 = 0$

$x = 1$

Substituting $x = 1$ in the given equation $y = 2x^2 - 4x$ yields $y = -2$; thus $(1,-2)$ is the critical point.

Step 2. Test of critical point: Since $f''(x) = 4$ for all x,

$f''(1) = 4 > 0$

Thus $(1,-2)$ is a minimum.

Step 3. Concavity: Since $f''(x) = 4 > 0$, the graph is concave up everywhere.

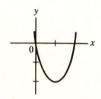

5. $f(x) = -4 - 3x - \frac{1}{2}x^2$; $f'(x) = -3 - x$; $f''(x) = -1$

Step 1. Critical points:

$f'(x) = -3 - x = 0$

$x = -3$

Substituting $x = -3$ in $y = -4 - 3x - \frac{1}{2}x^2$, we get $y = \frac{1}{2}$. So $(-3,\frac{1}{2})$ is the critical point.

Step 2. Test of critical point: Since $f''(x) = -1$ for all x,

$f''(-3) = -1 < 0$

So $(-3,\frac{1}{2})$ is a maximum.

<u>Step 3</u>. Concavity: Since $f''(x) = -1$, the graph is concave down
everywhere.

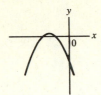

9. $f(x) = x^3 - 6x^2 + 9x - 3$; $f'(x) = 3x^2 - 12x + 9$;
$f''(x) = 6x - 12$

<u>Step 1</u>. Critical points:
$$f'(x) = 3x^2 - 12x + 9$$
$$3(x^2 - 4x + 3) = 0$$
$$3(x - 3)(x - 1) = 0$$
$$x = 1, 3$$
Substituting these values in $y = x^3 - 6x^2 + 9x - 3$, we see
that $(1,1)$ and $(3,-3)$ are the critical points.

<u>Step 2</u>. Test of critical points:
$$f''(1) = 6 - 12 = -6 < 0; \text{ thus } (1,1) \text{ is a maximum}$$
$$f''(3) = 18 - 12 = 6 > 0; \text{ thus } (3,-3) \text{ is a minimum}$$

<u>Step 3</u>. Concavity: We need to determine where $y'' < 0$ and where
$y'' > 0$. To this end, we first determine where $y'' = 0$:
$$f''(x) = 6x - 12 = 0 \text{ when } x = 2$$
If $x = 2$, then $y = -1$; the point is $(2,-1)$.

	test values	$y'' = 6x - 12$
$x < 2$	1	-
$x > 2$	3	+

So $f(x)$ is concave down to the left of $(2,-1)$ and $f(x)$
is concave up to the right of $(2,-1)$.

<u>Step 4</u>. Inflection point: Since the concavity changes, the point
$(2,-1)$ is an inflection point).

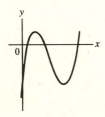

13. $f(x) = x^2 - 6x$; $f'(x) = 2x - 6$; $f''(x) = 2$

Step 1. Critical points: $2x - 6 = 0$ when $x = 3$. Substituting $x = 3$ in $y = x^2 - 6x$, we get $y = -9$. Thus $(3, -9)$ is the critical point.

Step 2. Test of critical point: $f''(x) = 2$ (minimum).

Step 3. Concavity: Since $f''(x) = 2$, the graph is concave up everywhere.

17. $f(x) = x^4 + x^3$; $f'(x) = 4x^3 + 3x^2$; $f''(x) = 12x^2 + 6x$

Step 1. Critical points:

$$4x^3 + 3x^2 = 0$$
$$x^2(4x + 3) = 0$$
$$x = -\frac{3}{4},\ 0$$

Substituting in $y = x^4 + x^3$, we get $\left(-\frac{3}{4}, -\frac{27}{256}\right)$ and $(0, 0)$ for the critical points.

Step 2. Test of critical points:

$f''\left(-\frac{3}{4}\right) > 0$; the point is a minimum

$f''(0) = 0$; the test fails

Using the first derivative test, if $x = -\frac{1}{2}$, $y' > 0$; if $x = 1$, $y' > 0$. So $(0, 0)$ is neither a minimum nor a maximum.

Step 3. Concavity: We need to determine where $y'' < 0$ and where $y'' > 0$. To this end we find those values of x for which $f''(x) = 0$:

$$f''(x) = 12x^2 + 6x = 0$$
$$6x(2x + 1) = 0$$
$$x = -\frac{1}{2},\ 0$$

Substituting in $y = x^4 + x^3$, we find that the points are $\left(-\frac{1}{2}, -\frac{1}{16}\right)$ and $(0, 0)$.

	test values	$6x$	$2x + 1$	$y'' = 6x(2x + 1)$
$x < -1/2$	-1	$-$	$-$	$+$
$-1/2 < x < 0$	$-1/4$	$-$	$+$	$-$
$x > 0$	1	$+$	$+$	$+$

Summary:

$f(x)$ is concave up for $x < -\frac{1}{2}$

$f(x)$ is concave down for $-\frac{1}{2} < x < 0$

$f(x)$ is concave up for $x > 0$

Step 4. Inflection points: Since the concavity changes at each of the points, $\left(-\frac{1}{2}, -\frac{1}{16}\right)$ and $(0,0)$ are points of inflection.

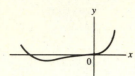

21. $f(x) = \dfrac{x}{x - 3}$; $f'(x) = -\dfrac{3}{(x - 2)^2}$; $f''(x) = \dfrac{6}{(x - 3)^3}$

Step 1. Critical points:
$$f'(x) = -\frac{3}{(x - 3)^2} < 0, \quad (x \neq 3)$$
Since $f'(x) \neq 0$, there are no critical points. In fact, the graph is strictly decreasing. [At $x = 3$, neither $f'(x)$ nor $f(x)$ is defined.]

Step 2. (Does not apply.)

Step 3. Concavity and points of inflection: Since $f''(x) \neq 0$, there are no inflection points. [Since $f''(3)$ does not exist, $x = 3$ is a possible point of inflection. However, since $f(x)$ is not defined at $x = 3$, this point cannot be an inflection point either.]
$$\frac{6}{(x - 3)^3} > 0 \text{ for } x > 3 \text{ (concave up)}$$
$$\frac{6}{(x - 3)^3} < 0 \text{ for } x < 3 \text{ (concave down)}$$

Step 4. Other: Since the denominator of $f(x) = \dfrac{x}{x - 3}$ is 0 when $x = 3$, the line $x = 3$ is a vertical asymptote. Also, since
$$\lim_{x \to \infty} \frac{x}{x - 3} = \lim_{x \to \infty} \frac{1}{1 - 3/x} = 1$$
$y = 1$ is a horizontal asymptote. (Note that the concavity changes at the vertical asymptote.)

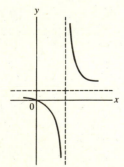

23. $f(x) = x^2 + \frac{8}{x}$; $f'(x) = 2x - \frac{8}{x^2}$; $f''(x) = 2 + \frac{16}{x^3}$

Step 1. Critical points:

$$f'(x) = 2x - \frac{8}{x^2} = \frac{2x^3 - 8}{x^2} = 0, \text{ or } x = \sqrt[3]{4}$$

Step 2. Test of critical points:

$$f''(\sqrt[3]{4}) > 0 \text{ (minimum)}$$

Step 3. Concavity: We need to determine where $y'' < 0$ and $y'' > 0$. To this end we first determine those values of x for which $f''(x) = 0$:

$$f''(x) = 2 + \frac{16}{x^3} = \frac{2x^3 + 16}{x^3} = 0$$

$$2x^3 + 16 = 0$$

$$x^3 = -8$$

$$x = -2$$

From $y = x^2 + \frac{8}{x}$, the point is $(-2,0)$.

If $x = -3$, $y'' > 0$; so $f(x)$ is concave up for $x < -2$.

If $x = -1$, $y'' < 0$; so $f(x)$ is concave down for $-2 < x < 0$. We see that the concavity changes at $(-2,0)$. Since $y'' > 0$ for $x > 0$, the graph is also concave up for $x > 0$.

Step 4. Inflection point: Since the concavity changes, the point $(-2,0)$ is an inflection point.

Step 5. Other: If x gets large,

$$y = x^2 + \frac{8}{x}$$

approaches $y = x^2$, which is therefore an asymptotic curve. Since the denominator of $f(x)$ is 0 when $x = 0$, the y-axis is a vertical asymptote. (Note that the concavity changes at the asymptote.)

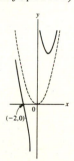

25. $f(x) = \frac{x + 1}{x - 2}$

   Vertical asymptote: $x = 2$

   Horizontal asymptote: $y = 1$, since

$$\lim_{x \to \infty} \frac{x + 1}{x - 2} = \lim_{x \to \infty} \frac{1 + \frac{1}{x}}{1 - \frac{2}{x}} = 1$$

$$f'(x) = \frac{-3}{(x - 2)^2}; \quad f''(x) = \frac{6}{(x - 2)^3}$$

Step 1. Critical points: Since

$$f'(x) = \frac{-3}{(x - 2)^2} < 0 \text{ for all } x \ne 2,$$

there are no critical points. Moreover, $f(x)$ is always decreasing.

Step 2. Does not apply.

Step 3. Concavity:

$$f''(x) = \frac{6}{(x - 2)^3}$$

$f''(x) > 0$ for $x > 2$ (concave up)

$f''(x) < 0$ for $x < 2$ (concave down)

Step 4. Inflection points: none
(Note that the concavity changes at $x = 2$, the vertical asymptote.)

29. $f(x) = (x - 3)^{1/3}; \quad f'(x) = \frac{1}{3(x - 3)^{2/3}}; \quad f''(x) = -\frac{2}{9(x - 3)^{5/3}}$

Neither $f'(x)$ nor $f''(x)$ are 0 for any value of x, but $f'(3)$ and $f''(3)$ do not exist. Thus $(3,0)$ is a critical point. Since $f'(x) > 0$ for all $x \ne 3$, $f(x)$ is strictly increasing. Also, $f''(x) > 0$ for $x < 3$ and $f''(x) < 0$ for $x > 3$. So $(3,0)$ is an inflection point. Finally, since $f'(x)$ approaches infinity when x approaches 3, there is a vertical tangent line at $(3,0)$.

33. $f(x) = \frac{6x}{x^2 + 3}; \quad f'(x) = \frac{6(3 - x^2)}{(x^2 + 3)^2}; \quad f''(x) = \frac{12x(x^2 - 9)}{(x^2 + 3)^3}$

Step 1. Critical points:
$$6(3 - x^2) = 0 \text{ when } x = -\sqrt{3}, \sqrt{3}$$

Step 2. Test of critical points:

$f''(-\sqrt{3}) > 0$ (minimum)

$f''(\sqrt{3}) < 0$ (maximum)

Step 3. Concavity: We need to determine where $y'' < 0$ and where $y'' > 0$. To this end we first determine where $y'' = 0$:

$$12x(x^2 - 9) = 0$$
$$x = 0, \pm 3$$

	test values	$12x$	$x^2 - 9$	$12x(x^2 - 9)/(x^2 + 3)^3$
$x < -3$	-4	$-$	$+$	$-$
$-3 < x < 0$	-1	$-$	$-$	$+$
$0 < x < 3$	1	$+$	$-$	$-$
$x > 3$	4	$+$	$+$	$+$

Summary:

$f(x)$ is concave down for $x < -3$

$f(x)$ is concave up for $-3 < x < 0$

$f(x)$ is concave down for $0 < x < 3$

$f(x)$ is concave up for $x > 3$

Step 4. Inflection points: Since the concavity changes, we get inflection points at $x = 0$ and $x = \pm 3$

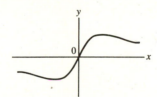

Section 3.4

1. $y = x^5 + x^3 + 1$

$y' = 5x^4 + 3x^2 = 0$ $y'' = 20x^3 + 6x = 0$

$x^2(5x^2 + 3) = 0$ $x(20x^2 + 6) = 0$

$x = 0, \; 5x^2 + 3 > 0$ $x = 0, \; 20x^2 + 6 > 0$

Critical point: $(0,1)$ inflection point: $(0,1)$

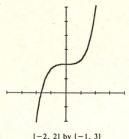

[-2, 2] by [-1, 3]

5. $y = 3.0x^2 + 1.2x^5 - 1.0$

$y' = 6x + 1.2(5x^4) = 0$ 　　　　　　 $y'' = 6 + 24x^3 = 0$

　　　　$6x + 6x^4 = 0$ 　　　　　　 $6(1 + 4x^3) = 0$

　　　$6x(1 + x^3) = 0$ 　　　　　　 $4x^3 = -1$

　　　　　　$x = 0, -1$ 　　　　　　 $x^3 = -\frac{1}{4}$

　　　　　　　　　　　　　　　　　 $x = -\frac{1}{\sqrt[3]{4}} \approx -0.63$

If $x = 0$, $y = -1.0$

If $x = -1$, $y = 3.0(-1)^2 + 1.2(-1)^5 - 1.0$

　　　　　　$= 0.80$

If $x = -\frac{1}{\sqrt[3]{4}}$, $y = 3.0\left(-\frac{1}{\sqrt[3]{4}}\right)^2 + 1.2\left(-\frac{1}{\sqrt[3]{4}}\right)^5 - 1.0$

　　　　　　　$= 0.071$

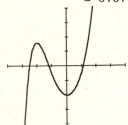

[-2, 2] by [-2, 2]

9. $y = 1.5x^4 - 0.50x^6 + 0.20$

$y' = 1.5(4x^3) - 0.50(6x^5) = 0$ 　　　 $y'' = 18x^2 - 15x^4 = 0$

　　　　　　$6x^3 - 3x^5 = 0$ 　　　　　 $x^2(18 - 15x^2) = 0$

　　　　　$3x^3(2 - x^2) = 0$ 　　　　 $x = 0, \ x^2 = \frac{18}{15}$

　　　　$x = 0, \pm\sqrt{2}$

　　　　　　　　　　　　　　　　 $x = \pm\sqrt{\frac{18}{15}}$

　　　　　　　　　　　　　　　　 $\approx \pm1.1$

If $x = 0$, $y = 0.20$

If $x = \pm\sqrt{2}$, $y = 1.5\left(\pm\sqrt{2}\right)^4 - 0.50\left(\pm\sqrt{2}\right)^6 + 0.20$

$$= 2.2$$

If $x = \pm\sqrt{\frac{18}{15}}$, $y = 1.5\left(\pm\sqrt{\frac{18}{15}}\right)^4 - 0.50\left(\pm\sqrt{\frac{18}{15}}\right)^6 + 0.20$

$$= 1.5$$

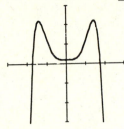

[-3, 3] by [-3, 3]

13.　$y = \frac{1}{7}x^7 - \frac{1}{5}x^5$

$y' = x^6 - x^4 = 0$ 　　　　　　　　　　$y'' = 6x^5 - 4x^3 = 0$

$\quad x^4(x^2 - 1) = 0$ 　　　　　　　　　$x^3(6x^2 - 4) = 0$

$\quad x = 0,\ \pm 1$ 　　　　　　　　　　$x = 0,\ x^2 = \frac{2}{3}$

$$x = \pm\frac{\sqrt{6}}{3}$$

If $x = 0$, $y = 0$

If $x = 1$, $y = \frac{1}{7} - \frac{1}{5} = -\frac{2}{35}$

If $x = -1$, $y = -\frac{1}{7} + \frac{1}{5} = \frac{2}{35}$

If $x = \frac{\sqrt{6}}{3}$, $y = \frac{1}{7}\left(\frac{\sqrt{6}}{3}\right)^7 - \frac{1}{5}\left(\frac{\sqrt{6}}{3}\right)^5 \approx -0.038$

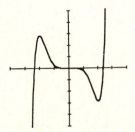

[-2, 2] by [-0.1, 0.1]

59

17. By the quotient rule,

$$\frac{dy}{dx} = \frac{(x^2 - 1) \cdot 1 - x(2x)}{(x^2 - 1)^2} = \frac{x^2 - 1 - 2x^2}{(x^2 - 1)^2}$$

$$= \frac{-x^2 - 1}{(x^2 - 1)^2} = - \frac{x^2 + 1}{(x^2 - 1)^2} \neq 0$$

In fact, $y' < 0$ for all x, so that the slope of the tangent line is negative everywhere. (See graph in answer section.)

21. $y = 0.50x^3 + 2.34x^{-1}$

$y' = 0.50(3x^2) - 2.34x^{-2}$

$\quad = 1.50x^2 - \frac{2.34}{x^2} = 0$

Multiplying both sides of the equation by x^2, we get

$\quad 1.50x^4 - 2.34 = 0$

$$x = \pm \sqrt[4]{\frac{2.34}{1.50}} \approx \pm 1.12$$

If $x = 1.12$, $y = 0.50(1.12)^3 + \frac{2.34}{1.12} = 2.79$

If $x = -1.12$, $y = -2.79$

(See graph in answer section.)

Section 3.5

1. Find the critical point:

$\frac{dP}{di} = 0 + 12.8 - 6.40i = 0$

$i = \frac{12.8}{6.40} = 2.0$ A

Since

$$\frac{d^2P}{di^2} = -6.40 < 0,$$

$i = 2.0$ leads to a maximum. Thus
$P = 4.50 + 12.8(2.0) - 3.20(2.0)^2 = 17.3$ W

5. $C = k_1A + k_2A^{-1}$

$\frac{dC}{dA} = k_1 + k_2(-1A^{-2})$ since k_1 and k_2 are constants

$\frac{dC}{dA} = 0$ when

$\quad k_1 + k_2(-A^{-2}) = 0$

$\quad\quad k_1 - \frac{k_2}{A^2} = 0$

$\quad\quad k_1A^2 - k_2 = 0$ multiplying by A^2

$$A^2 = \frac{k_2}{k_1}$$

$$A = \sqrt{\frac{k_2}{k_1}}$$

Since $\dfrac{d^2C}{dA^2} = k_2(2A^{-3}) = \dfrac{2k_2}{A^3} > 0$ for $A > 0$, C is a minimum at the critical value.

9. Let $M(x) =$ the marginal cost, that is,
$$M(x) = C'(x) = 3(2.0 \times 10^{-6})x^2 - 0.0030x + 2.5$$
$$= 6.0 \times 10^{-6}x^2 - 0.0030x + 2.5$$

Then
$$M'(x) = 2(6.0 \times 10^{-6})x - 0.0030 = 0$$
$$x = \frac{0.0030}{2(6.0 \times 10^{-6})} = 250 \text{ units}$$

Since $M''(x) = 2(6.0 \times 10^{-6}) > 0$, $x = 250$ corresponds to the minimum.

13. Referring to Figure 3.23, the quantity to be minimized is
$$L = x + 2y$$
To eliminate one of the variables, we use the fact that the area is 200 m^2:
$$xy = 200 \text{ or } x = \frac{200}{y}$$

Thus
$$L = \frac{200}{y} + 2y = 200y^{-1} + 2y$$

$$L' = -200y^{-2} + 2 = 0$$

Multiplying by y^2, we get
$$-200 + 2y^2 = 0$$
$$2y^2 = 200 \text{ and } y = 10$$
$$y = 10 \text{ m}$$
$$x = \frac{200}{10} = 20 \text{ m}$$

From $L'' = 400y^{-3} = \dfrac{400}{y^3}$, we get $L'' = \dfrac{400}{10^3} > 0$, so that $y = 10$ m leads to the minimum value for L.

17.

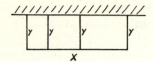

Quantity to be minimized: $L = x + 4y$
Eliminate x: $xy = 1600$ or $x = \frac{1600}{y}$

$L = \frac{1600}{y} + 4y = 1600y^{-1} + 4y$

$L' = -1600y^{-2} + 4 = 0$

$-400y^{-2} + 1 = 0$

After multiplying by y^2, we get $-400 + y^2 = 0$, or $y = 20$ ft.

21. Let x = width and y = depth. Then S, the strength of the beam, is
$$S = kxy^2, \text{ k a constant.}$$
This is the quantity to be maximized. To eliminate y, note that from the Pythagorean theorem,

$x^2 + y^2 = d^2$
$\qquad y^2 = 9 - x^2 \qquad d = 3$ ft

So

$S = kx(9 - x^2)$
$\quad = k(9x - x^3)$
$S' = k(9 - 3x^2) = 0$
$\qquad 9 - 3x^2 = 0$
$\qquad\qquad x^2 = 3$
$\qquad\qquad y^2 = 9 - 3 = 6$

So $x = \sqrt{3}$ ft and $y = \sqrt{6}$ ft.

25.

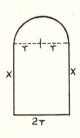

Area of rectangle: $2rx$

Area of semicircle: $\frac{1}{2}\pi r^2$

62

Quantity to be maximized: $A = 2rx + \frac{1}{2}\pi r^2$. To eliminate x, note that the perimeter is

$$5 = 2x + 2r + \pi r$$
$$x = \frac{1}{2}(5 - 2r - \pi r)$$

Substituting in A, we get

$$A = 2r \cdot \frac{1}{2}(5 - 2r - \pi r) + \frac{1}{2}\pi r^2$$
$$= 5r - 2r^2 - \pi r^2 + \frac{1}{2}\pi r^2$$
$$= 5r - 2r^2 - \frac{1}{2}\pi r^2$$
$$A' = 5 - 4r - \pi r = 0$$
$$(4 + \pi)r = 5$$
$$r = \frac{5}{4 + \pi} \text{ m}$$

29. Let (x,y) be a point on the curve. Then
$$d^2 = (x - 1)^2 + (y - 2)^2$$
Since $y = \frac{1}{4}x^2$,
$$d^2 = (x - 1)^2 + \left(\frac{1}{4}x^2 - 2\right)^2$$
$$= x^2 - 2x + 1 + \frac{1}{16}x^4 - x^2 + 4$$
$$= \frac{1}{16}x^4 - 2x + 5$$

Critical value:
$$\frac{1}{16}(4x^3) - 2 = 0$$
$$4x^3 = 32$$
$$x^3 = 8 \text{ and } x = 2$$
$$y = \frac{1}{4}(2)^2 = 1$$

33. Let x = the number of passengers above 30.

Then 400 - 10x = price per ticket.

Intake: I = price per ticket times the number of passengers, which is 30 + x.

$$I = (400 - 10x)(30 + x)$$
$$= -10x^2 + 100x + 12{,}000$$
$$I' = -20x + 100 = 0 \quad \text{or} \quad x = 5$$

So 30 + x = 35 passengers

37. Let h and r be the height and radius of the cylinder and h_1 and r_1 the height and radius of the cone, so that h_1 and r_1 are fixed quantities.

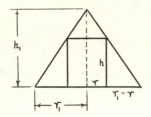

Quantity to be maximized: $V = \pi r^2 h$. To eliminate h, note that by similar triangles we get

$$\frac{h}{h_1} = \frac{r_1 - r}{r_1} \quad \text{or} \quad h = \frac{h_1(r_1 - r)}{r_1} \tag{1}$$

Substituting in V:

$$V = \pi r^2 \frac{h_1(r_1 - r)}{r_1}$$

$$= \frac{\pi h_1}{r_1}(r_1 r^2 - r^3), \quad h_1 \text{ and } r_1 \text{ constants}$$

$$V' = \frac{\pi h_1}{r_1}(2r_1 r - 3r^2) = 0$$

$$r = \frac{2}{3}r_1$$

By (1)

$$h = \frac{h_1}{r_1}\left(r_1 - \frac{2}{3}r_1\right) = \frac{1}{3}h_1$$

So the height of the cylinder is one-third the height of the cone.

Section 3.6

1. $\frac{dy}{dt} = 2x\frac{dx}{dt} = 2(2)(1) = 4$

5. $I = \frac{E}{R} = \frac{100}{R} = 100R^{-1}$. Given $\frac{dR}{dt} = 2$, find $\frac{dI}{dt}$ when $R = 10$.

$\frac{dI}{dt} = 100(-1)R^{-2}\frac{dR}{dt} = -\frac{100}{R^2}\frac{dR}{dt} = -\frac{(100)(2)}{R^2}\Big|_{R=10} = -2 \text{ A/s}$

9. In the diagram we label all quantities that change and all quantities that remain fixed.

Given: $\frac{dx}{dt} = 2$

Find: $\frac{dy}{dt}$ when $x = 4$

$$x^2 + y^2 = 25$$

$$2x\frac{dx}{dt} + 2y\frac{dy}{dt} = 0$$

$$x\frac{dx}{dt} + y\frac{dy}{dt} = 0$$

$$2x + y\frac{dy}{dt} = 0 \qquad \frac{dx}{dt} = 2$$

$$\frac{dy}{dt} = -\frac{2x}{y}$$

At the instant when $x = 4$, we have $y = 3$. Thus

$$\frac{dy}{dt} = -\frac{(2)(4)}{3} = -\frac{8}{3} \text{ m/min}$$

13. First we need to compute k from the formula $PV^{1.4} = k$. At the instant in question, $V = 2.0$ and $P = 76$. Thus

$$k = 76(2.0)^{1.4}$$

Given $\frac{dV}{dt} = -1.0$, find $\frac{dP}{dt}$ when $V = 2.0$. From $PV^{1.4} = k$, we have

$$P = kV^{-1.4}$$

$$\begin{aligned}
\frac{dP}{dt} &= k(-1.4)V^{-2.4}\frac{dV}{dt} \\
&= k(-1.4)V^{-2.4}(-1.0) = \frac{1.4k}{V^{2.4}}\Big|_{V=2.0} \\
&= \frac{(1.4)(76)(2.0)^{1.4}}{(2.0)^{2.4}} \qquad k = 76(2.0)^{1.4} \\
&= 53 \text{ Pa/min}
\end{aligned}$$

17. In the diagram, we label all quantities that change and all quantities that remain fixed.

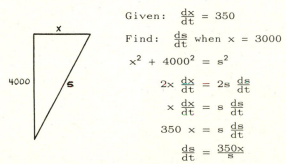

Given: $\frac{dx}{dt} = 350$

Find: $\frac{ds}{dt}$ when $x = 3000$

$$x^2 + 4000^2 = s^2$$

$$2x\frac{dx}{dt} = 2s\frac{ds}{dt}$$

$$x\frac{dx}{dt} = s\frac{ds}{dt}$$

$$350 x = s\frac{ds}{dt}$$

$$\frac{ds}{dt} = \frac{350x}{s}$$

At the instant when $x = 3000$, we have $s = 5000$. Thus

$$\frac{ds}{dt} = \frac{(350)(3000)}{5000} = \frac{(350)(3)}{5} = (70)(3) = 210 \text{ km/h}$$

21. Given $\frac{dx}{dt} = -2$, find $\frac{dy}{dt}$ when $x = 16$ and $y = 4$. From $y^2 = x$, we have

$$2y\frac{dy}{dt} = \frac{dx}{dt} = -2$$

$$\frac{dy}{dt} = -\frac{1}{y}\Big|_{y=4} = -\frac{1}{4}\frac{\text{unit}}{\text{min}}$$

65

25.

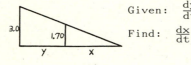

Since $h = \frac{3}{4}r$, $r = \frac{4}{3}h$

$V = \frac{1}{3}\pi r^2 h = \frac{1}{3}\pi\left(\frac{4}{3}h\right)^2 h$

$V = \frac{16\pi}{27}h^3$

Given $\frac{dV}{dt} = 12$, find $\frac{dh}{dt}$ when $h = 6.0$.

$\frac{dV}{dt} = \frac{16\pi}{27}(3h^2)\frac{dh}{dt}$

$12 = \frac{16\pi}{9}h^2\frac{dh}{dt}$ $\qquad\qquad \frac{dV}{dt} = 12$

$\frac{dh}{dt} = \frac{108}{16\pi h^2}\Big|_{h=6.0} = \frac{108}{16\pi(36)} = \frac{3}{16\pi} = 0.060$ ft/s

29.

Given: $\frac{dy}{dt} = 2.0$

Find: $\frac{dx}{dt}$

By similar triangles:

$\frac{x + y}{3.0} = \frac{x}{1.70}$

$x + y = \frac{3.0}{1.70}x$

$y = \frac{1.3}{1.70}x$

$\frac{dy}{dt} = \frac{1.3}{1.70} \cdot \frac{dx}{dt}$

$2.0 = \frac{1.3}{1.70} \cdot \frac{dx}{dt}$

$\frac{dx}{dt} = 2.6$ m/s

33.

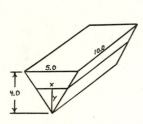

Given: $\frac{dV}{dt} = 2.0$

Find: $\frac{dy}{dt}$ when $y = 3.0$

By similar triangles:

$\frac{x}{y} = \frac{5.0}{4.0}$ or $x = \frac{5}{4}y$

$$\text{Volume} = \text{Base} \times \text{height} = \tfrac{1}{2}xy \times 10.0$$
$$V = 5.0xy$$
$$V = 5.0\left(\tfrac{5}{4}y\right)y \quad\text{since}\quad x = \tfrac{5}{4}y$$
$$V = \tfrac{25}{4}y^2$$
$$\frac{dV}{dt} = \frac{25}{4}(2y)\frac{dy}{dt}$$
$$\frac{dV}{dt} = \frac{25y}{2}\frac{dy}{dt}$$
$$2.0 = \frac{25y}{2}\frac{dy}{dt} \quad\text{since}\quad \frac{dV}{dt} = 2.0$$
$$\frac{dy}{dt} = \left.\frac{4.0}{25y}\right|_{y=3.0} = \frac{4}{75} = 0.053 \text{ m/min}$$

Section 3.7

1. Since $\frac{dy}{dx} = 3x^2 - 1$, $dy = (3x^2 - 1)dx$

5. If $x = 2$, $y = 2^2 - 2 = 2$. If $x = 2.1$, $y = (2.1)^2 - 2.1 = 2.31$.
   Hence $\Delta y = 2.31 - 2 = 0.31$.
   From $y = x^2 - x$, we get the differential
   $dy = (2x - 1)dx$.
   So if $x = 2$ and $dx = 0.1$, $dy = (4 - 1)(0.1) = 0.3$.

9. $V = s^3$, where s is the length of a side
   $s = 5.00$ in. and $ds = \pm 0.01$ in.
   $\Delta V \approx dV = 3s^2 ds = 3(5.00)^2(\pm 0.01) = \pm 0.75$ in.^{3}
   Percentage error: $\frac{dV}{V} \times 100 = \frac{0.75}{(5.00)^3} \times 100 = 0.6\%$

13. $T = 2\pi\sqrt{\frac{L}{10}} = \frac{2\pi}{\sqrt{10}}L^{1/2}$
    $L = 2.0$ m and $dL = \pm 0.1$ m
    $$\Delta T \approx dT = \frac{2\pi}{\sqrt{10}} \cdot \tfrac{1}{2}L^{-1/2}dL$$
    $$= \frac{\pi}{\sqrt{10}}\frac{1}{\sqrt{L}}dL$$
    $$= \frac{\pi}{\sqrt{10}}\frac{1}{\sqrt{2.0}}(\pm 0.1) = \pm 0.07 \text{ s}$$

    Percentage error: $\frac{dT}{T} \times 100 = \frac{0.07}{2\pi\sqrt{\frac{2.0}{10}}} \times 100 = 2.5\%$

17. $A = \pi r^2$

$\Delta A \approx dA = 2\pi r dr$

The geometric interpretation of this formula can be seen from the following diagram:

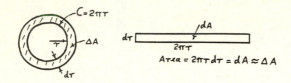

REVIEW EXERCISES FOR CHAPTER 3

1. $y = (x - 2)^{1/2}$, $y' = \frac{1}{2}(x - 2)^{-1/2} = \frac{1}{2\sqrt{x - 2}}\Big|_{x=3} = \frac{1}{2}$

Tangent line: $y - 1 = \frac{1}{2}(x - 3)$ $m = \frac{1}{2}$

$2y - 2 = x - 3$

$x - 2y - 1 = 0$

Normal line: $y - 1 = -2(x - 3)$ $m = \frac{-1}{1/2}$

$y - 1 = -2x + 6$

$2x + y - 7 = 0$

5. $f(x) = -x^3 + 12x + 2$; $f'(x) = -3x^2 + 12$; $f''(x) = -6x$

Step 1. Critical points:

$f'(x) = -3x^2 + 12 = 0$

$x = \pm 2$

The points are $(2,18)$ and $(-2,-14)$.

Step 2. Test of critical points:

$f''(2) = -12 < 0$ (maximum)

$f''(-2) = 12 > 0$ (minimum)

Step 3. Concavity:

$f''(x) = -6x = 0$ when $x = 0$.

If $x < 0$, $y'' > 0$ (concave up).

If $x > 0$, $y'' < 0$ (concave down).

Step 4. Inflection point: Since the concavity changes, the point $(0,2)$ is a point of inflection.

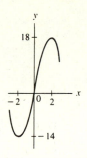

9. $f(x) = x^2 - \frac{1}{x}$; $f'(x) = 2x + \frac{1}{x^2}$; $f''(x) = 2 - \frac{2}{x^3}$

Step 1. Critical points:
$$f'(x) = 2x + \frac{1}{x^2} = 0$$
$$2x^3 + 1 = 0$$
$$2x^3 = -1$$
$$x = - \frac{1}{\sqrt[3]{2}}$$

Step 2. Test of critical points:
$$f''(-\frac{1}{\sqrt[3]{2}}) > 0 \quad (minimum)$$

Step 3. Concavity: We need to determine where $y'' < 0$ and where $y'' > 0$. To this end we first determine those values of x for which $f''(x) = 0$:

$$f''(x) = 2 - \frac{2}{x^3} = 0$$
$$2x^3 - 2 = 0$$
$$x = 1$$

(The concavity may also change at the vertical asymptote $x = 0$.)

	test values	$y'' = 2 - \frac{2}{x^3}$
$x > 1$	2	+
$0 < x < 1$	1/2	−
$x < 0$	−1	+

Summary:

If $x > 1$, $f(x)$ is concave up

If $0 < x < 1$, $f(x)$ is concave down

If $x < 0$, $f(x)$ is concave up

<u>Step 4.</u> Inflection point: Because of the change in concavity, (1,0) is a point of inflection. (Note that the concavity also changes at x = 0, the vertical asymptote.)

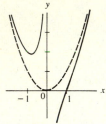

13.

Denote the radius of the log by r, the width of the beam by x, and the depth by y.

Quantity to be maximized: $S = kxy^3$. To eliminate x, note that

$$x^2 + y^2 = 4r^2$$
$$x = \sqrt{4r^2 - y^2}$$

Substituting in the expression for S, we get

$$S = ky^3(4r^2 - y^2)^{1/2}$$
$$S' = ky^3 \cdot \frac{1}{2}(4r^2 - y^2)^{-1/2}(-2y) + k(3y^2)(4r^2 - y^2)^{1/2}$$
$$= k\left[\frac{-y^4}{\sqrt{4r^2 - y^2}} + 3y^2\sqrt{4r^2 - y^2}\right] = 0$$

Multiplying by $\frac{1}{k}\sqrt{4r^2 - y^2}$, we get

$$-y^4 + 3y^2(4r^2 - y^2) = 0$$
$$-y^4 + 12r^2y^2 - 3y^4 = 0$$
$$12r^2y^2 - 4y^4 = 0$$
$$4y^2(3r^2 - y^2) = 0$$
$$y^2 = 3r^2$$
$$y = \sqrt{3}r$$

Hence $x = \sqrt{4r^2 - 3r^2} = r$. So the desired ratio is

$$\frac{y}{x} = \frac{\sqrt{3}r}{r} = \sqrt{3}$$

70

17.

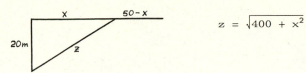

$$z = \sqrt{400 + x^2}$$

Quantity to be minimized:

$$C = 45\sqrt{400 + x^2} + 30(50 - x)$$

$$C' = 45 \cdot \tfrac{1}{2}(400 + x^2)^{-1/2}(2x) - 30 = 0$$

$$\frac{45x}{\sqrt{400 + x^2}} = 30$$

$$\frac{3x}{\sqrt{400 + x^2}} = 2$$

$$3x = 2\sqrt{400 + x^2}$$

$$9x^2 = 4(400 + x^2)$$

$$5x^2 = 1600$$

$$x^2 = 320 = (64)(5)$$

$$x = 8\sqrt{5}$$

So the distance on land is $(50 - 8\sqrt{5})$m.

21.

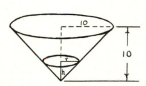

$$\frac{r}{h} = \frac{10}{10} = 1, \text{ or } r = h$$

$$V = \tfrac{1}{3}\pi r^2 h = \tfrac{1}{3}\pi h^3$$

Given $\dfrac{dV}{dt} = 9$, find $\dfrac{dh}{dt}$ when $h = 3$.

$$\frac{dV}{dt} = \pi h^2 \frac{dh}{dt}$$

$$9 = \pi h^2 \frac{dh}{dt}$$

$$\frac{dh}{dt} = \frac{9}{\pi h^2}\Big|_{h=3} = \frac{1}{\pi} \frac{m^3}{min}$$

25. $V = s^3$, $s = 10.00$ cm, $ds = \pm 0.02$ cm

$$\Delta V \approx dV = 3s^2 ds = 3(10.00)^2(\pm 0.02) = \pm 6 \text{ cm}^3.$$

Percentage error: $\dfrac{dV}{V} \times 100 = \dfrac{6}{(10.00)^3} \times 100 = 0.6\%$

CHAPTER 4   THE INTEGRAL

Section 4.1

1. By (4.3), $F(x) = 3x + C$

5. $f(x) = 2x^3 - 3x^2 + x$. By (4.2)

   $F(x) = 2\frac{x^4}{4} - 3\frac{x^3}{3} + \frac{x^2}{2} + C = \frac{1}{2}x^4 - x^3 + \frac{1}{2}x^2 + C$

11. $f(x) = \frac{1}{x^2} - 2 = x^{-2} - 2$. By (4.1) and (4.3)

   $F(x) = \frac{x^{-1}}{-1} - 2x + C = -\frac{1}{x} - 2x + C$

13. $f(x) = \frac{3}{x^2} + \frac{2}{\sqrt[3]{x}} = 3x^{-2} + 2x^{-1/3}$. By (4.2)

   $F(x) = 3\frac{x^{-1}}{-1} + 2\frac{x^{2/3}}{2/3} + C = -\frac{3}{x} + 3x^{2/3} + C$

Section 4.2

1. Subdivide the interval $[0,1]$ into n equal parts, each of length $1/n$. Thus $\Delta x_i = 1/n$.

   Choosing the right endpoint of each subinterval for x_i, we get

   $x_1 = 1 \cdot \frac{1}{n}, \; x_2 = 2 \cdot \frac{1}{n}, \; x_3 = 3 \cdot \frac{1}{n}, \; \ldots, \; x_i = i \cdot \frac{1}{n}, \; \ldots,$

   $x_n = n \cdot \frac{1}{n} = 1$

   Since $f(x) = x$, we get, for the corresponding altitudes,

   $f(x_1) = 1 \cdot \frac{1}{n}, \; f(x_2) = 2 \cdot \frac{1}{n}, \; \ldots, \; f(x_i) = i \cdot \frac{1}{n}, \; \ldots, \; f(x_n) = n \cdot \frac{1}{n}$

   Since $\Delta x_i = 1/n$, the sum of the areas is given by

   $\sum_{i=1}^{n} f(x_i) \Delta x_i = \sum_{i=1}^{n} i \cdot \frac{1}{n} \cdot \frac{1}{n} = \frac{1}{n^2} \sum_{i=1}^{n} i = \frac{1}{n^2} \cdot \frac{n(n + 1)}{2}$

   by formula A. So the exact area is

   $\int_0^1 x \, dx = \lim_{n \to \infty} \frac{n(n + 1)}{2n^2} = \lim_{n \to \infty} \frac{n^2 + n}{2n^2} = \lim_{n \to \infty} \frac{1 + 1/n}{2} = \frac{1}{2}$

5. Subdividing the interval $[0,2]$ into n equal parts, we get $\Delta x_i = \frac{2}{n}$.

   Choosing the right endpoint in each subinterval for x_i again, we have

   $x_1 = 1 \cdot \frac{2}{n}, \; x_2 = 2 \cdot \frac{2}{n}, \; \ldots, \; x_i = i \cdot \frac{2}{n}, \; \ldots, \; x_n = n \cdot \frac{2}{n} = 2$

72

For the altitudes, we get from $f(x) = 3x^2$

$f(x_1) = 3(1)^2 \frac{4}{n^2}$, $f(x_2) = 3(2)^2 \frac{4}{n^2}$, ..., $f(x_i) = 3(i)^2 \frac{4}{n^2}$,

$\qquad\qquad\qquad\qquad\qquad\qquad\qquad$ ..., $f(x_n) = 3(n)^2 \frac{4}{n^2}$

Since $\Delta x_i = \frac{2}{n}$, the sum of the areas is

$$\sum_{i=1}^{n} f(x_i)\Delta x_i = \sum_{i=1}^{n} 3i^2 \frac{4}{n^2} \cdot \frac{2}{n} = \frac{24}{n^3} \sum_{i=1}^{n} i^2$$

$$= \frac{24}{n^3} \cdot \frac{n(n+1)(2n+1)}{6} = \frac{4n(n+1)(2n+1)}{n^3}$$

by Formula B. Hence

$$\int_0^2 3x^2 \, dx = \lim_{n\to\infty} \frac{4n(n+1)(2n+1)}{n^3} = \lim_{n\to\infty} \frac{8n^3 + 12n^2 + 4n}{n^3}$$

$$= \lim_{n\to\infty} \frac{8 + 12/n + 4/n^2}{1} = 8$$

Section 4.3

5. $\int_0^1 (x^3 + 1)dx = \frac{1}{4}x^4 + x\big|_0^1 = \left(\frac{1}{4} + 1\right) - (0) = \frac{5}{4}$

9. $\int_1^3 \frac{1}{x^3} \, dx = \int_1^3 x^{-3}dx = -\frac{1}{2x^2}\big|_1^3 = -\frac{1}{18} + \frac{1}{2} = \frac{8}{18} = \frac{4}{9}$

Section 4.5

1. $\int \sqrt{x} \, dx = \int x^{1/2}dx = \frac{x^{3/2}}{3/2} + C = \frac{2}{3}x^{3/2} + C = \frac{2}{3}xx^{1/2} + C = \frac{2}{3}x\sqrt{x} + C$

5. $\int (2\sqrt{x} - 3x^2 + 1)dx = 2\frac{x^{3/2}}{3/2} - 3\frac{x^3}{3} + x + C = \frac{4}{3}x^{3/2} - x^3 + x + C$

9. $\int (2x^2 - 3)^3(4x)dx$ $\qquad$ Let $u = 2x^2 - 3$; then $du = 4xdx$

$\int u^3 du = \frac{1}{4}u^4 + C = \frac{1}{4}(2x^2 - 3)^4 + C$

13. (a) $\int (1 - x)dx = x - \frac{1}{2}x^2 + C$

(b) $\int (1 - x)^4 dx$ $\qquad$ Let $u = 1 - x$; then $du = -dx$

$\int (1 - x)^4 dx = -\int (1 - x)^4(-dx) = -\int u^4 du = -\frac{1}{5}u^5 + C$

$\qquad\qquad\qquad\qquad\qquad\qquad\qquad\qquad = -\frac{1}{5}(1 - x)^5 + C$

17. $\int (2x^2 + x)^3 (4x + 1)\,dx$ Let $u = 2x^2 + x$; then $du = (4x + 1)\,dx$

$\int u^3\,du = \frac{1}{4}u^4 + C = \frac{1}{4}(2x^2 + x)^4 + C$

21. $\int \frac{x\,dx}{\sqrt{1 - x^2}} = \int (1 - x^2)^{-1/2} x\,dx$ Let $u = 1 - x^2$; then $du = -2x\,dx$

$\int (1 - x^2)^{-1/2} x\,dx = -\frac{1}{2}\int (1 - x^2)^{-1/2}(-2x\,dx) = -\frac{1}{2}\int u^{-1/2}\,du$

$\qquad\qquad = -\frac{1}{2}\frac{u^{1/2}}{1/2} + C = -u^{1/2} + C = -\sqrt{1 - x^2} + C$

25. $\int \frac{\sqrt[4]{x}}{x}\,dx = \int x^{1/4}x^{-1}\,dx = \int x^{-3/4}\,dx = \frac{x^{1/4}}{1/4} + C = 4\sqrt[4]{x} + C$

29. $\int (x^2 + 1)^2\,dx$. If we try letting $u = x^2 + 1$, then $du = 2x\,dx$ and no substitution can be made. As noted in Example 4, the integral is not of the proper form. Consequently, we must multiply out the binomial and integrate term by term using (4.8):

$\int (x^2 + 1)^2\,dx = \int (x^4 + 2x^2 + 1)\,dx = \frac{1}{5}x^5 + \frac{2}{3}x^3 + x + C$

33. As in Exercise 29, we must multiply out the integrand:

$\int (1 + \sqrt{x})^2\,dx = \int (1 + 2x^{1/2} + x)\,dx = x + 2\frac{x^{3/2}}{3/2} + \frac{x^2}{2} + C$

$\qquad = x + \frac{4}{3}x^{3/2} + \frac{1}{2}x^2 + C$

37. $\int \frac{(2 - 4r)\,dr}{\sqrt[4]{r - r^2}} = \int (r - r^2)^{-1/4}(2 - 4r)\,dr$ Let $u = r - r^2$; then $du = (1 - 2r)\,dr$

$\int (r - r^2)^{-1/4}(2 - 4r)\,dr = \int (r - r^2)^{-1/4}(2)(1 - 2r)\,dr$

$\qquad\qquad = 2\int (r - r^2)^{-1/4}(1 - 2r)\,dr$

$\qquad\qquad = 2\int u^{-1/4}\,du = 2\frac{u^{3/4}}{3/4} + C$

$\qquad\qquad = \frac{8}{3}(r - r^2)^{3/4} + C$

41. $\int (x^3 + 1)^3 (5x^2)\,dx$ Let $u = x^3 + 1$; then $du = 3x^2\,dx$

$\int (x^3 + 1)^3 (5x^2)\,dx = 5\int (x^3 + 1)^3 x^2\,dx = \frac{5}{3}\int (x^3 + 1)^3 (3x^2)\,dx$

$\qquad\qquad = \frac{5}{3}\int u^3\,du = \frac{5}{3}\frac{u^4}{4} + C = \frac{5}{12}(x^3 + 1)^4 + C$

45. $\int (3 - 3x)\sqrt{4x - 2x^2}\, dx = \int (4x - 2x^2)^{1/2}(3 - 3x)dx$

Let $u = 4x - 2x^2$; then $du = (4 - 4x)dx = 4(1 - x)dx$

$\int (4x - 2x^2)^{1/2}(3)(1 - x)dx = 3\int (4x - 2x^2)^{1/2}(1 - x)dx$

$\qquad\qquad = \frac{3}{4}\int (4x - 2x^2)^{1/2}(4)(1 - x)dx$

$\qquad\qquad = \frac{3}{4}\int u^{1/2}du = \frac{3}{4}\frac{u^{3/2}}{3/2} + C$

$\qquad\qquad = \frac{1}{2}(4x - 2x^2)^{3/2} + C$

49. $\int (2x^3 + 1)^2(6x)dx$. If we let $u = 2x^3 + 1$, then $du = 6x^2 dx$, which is not the same as $6x\,dx$. So the substitution cannot be made and the expression in the integrand must be multiplied out instead:

$\int (2x^3 + 1)^2(6x)dx = \int (4x^6 + 4x^3 + 1)6x\,dx$

$= \int (24x^7 + 24x^4 + 6x)dx = 24\frac{x^8}{8} + 24\frac{x^5}{5} + 6\frac{x^2}{2} + C$

$= 3x^8 + \frac{24}{5}x^5 + 3x^2 + C$

53. As in Exercise 49, if $u = x^3 + 1$, then $du = 3x^2 dx$, which cannot be made to match $x\,dx$ in the integral. Instead, we multiply out the integrand and integrate term by term:

$\int (x^3 + 1)^2 x\,dx = \int (x^6 + 2x^3 + 1)x\,dx = \int (x^7 + 2x^4 + x)dx$

$\qquad\qquad = \frac{1}{8}x^8 + \frac{2}{5}x^5 + \frac{1}{2}x^2 + C$

57. $\int_0^1 (1 - x)dx = x - \frac{1}{2}x^2\Big|_0^1 = [1 - \frac{1}{2}(1)^2] - [0 - \frac{1}{2}(0)^2] = \frac{1}{2}$

61. $\int_0^1 \sqrt{1 - x}\, dx$

$u = 1 - x$, $du = -dx$

Lower limit: if $x = 0$, then $u = 1 - x = 1 - 0 = 1$

Upper limit: if $x = 1$, then $u = 1 - x = 1 - 1 = 0$

$\int_0^1 (1 - x)^{1/2}dx = -\int_0^1 (1 - x)^{1/2}(-dx) = -\int_1^0 u^{1/2}du$

$= -\frac{2}{3}u^{3/2}\Big|_1^0 = 0 - (-\frac{2}{3}) = \frac{2}{3}$

Alternatively, we can find the indefinite integral first and substitute $x = 0$ and $x = 1$:

$$\int_0^1 (1 - x)^{1/2}dx = -\int_0^1 (1 - x)^{1/2}(-dx) \qquad u = 1 - x; \ du = -dx$$

$$= -\tfrac{2}{3}(1 - x)^{3/2}\big|_0^1$$

$$= 0 - (-\tfrac{2}{3} \cdot 1) = \tfrac{2}{3}$$

65. $\displaystyle\int_2^7 \frac{dx}{\sqrt{x + 2}} = \int_2^7 (x + 2)^{-1/2}dx \qquad u = x + 2, \ du = dx$

Lower limit: if $x = 2$, then $u = x + 2 = 2 + 2 = 4$

Upper limit: if $x = 7$, then $u = x + 2 = 7 + 2 = 9$

$$\int_2^7 (x + 2)^{-1/2}dx = \int_4^9 u^{-1/2}du = 2u^{1/2}\big|_4^9 = 2\sqrt{u}\big|_4^9$$

$$= 2\sqrt{9} - 2\sqrt{4} = 6 - 4 = 2$$

Alternatively, we can find the indefinite integral first and
substitute $x = 2$ and $x = 7$:

$$\int_2^7 (x + 2)^{-1/2}dx = 2(x + 2)^{1/2}\big|_2^7 \qquad u = x + 2; \ du = dx$$

$$= 2(9)^{1/2} - 2(4)^{1/2} = 2$$

69. $\displaystyle\int_4^9 \frac{1 + \sqrt{r}}{\sqrt{r}}dr = \int_4^9 \frac{1 + r^{1/2}}{r^{1/2}}dr = \int_4^9\left(\frac{1}{r^{1/2}} + \frac{r^{1/2}}{r^{1/2}}\right)dr$

$$= \int_4^9 (r^{-1/2} + 1)dr = 2r^{1/2} + r\big|_4^9 = [2(9)^{1/2} + 9] - [2(4)^{1/2} + 4]$$

$$= (6 + 9) - (4 + 4) = 15 - 8 = 7$$

Section 4.6

1.

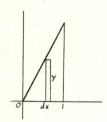

$$\int_0^1 y\,dx = \int_0^1 2x\,dx = x^2\big|_0^1 = 1$$

5.

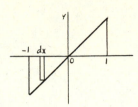

For the region on the left,

$$\int_{-1}^{0} (0 - x)\,dx = -\frac{x^2}{2}\Big|_{-1}^{0}$$

$$= 0 - \left(-\frac{1}{2}\right) = \frac{1}{2}$$

For the region on the right:

$$\int_{0}^{1} x\,dx = \frac{x^2}{2}\Big|_{0}^{1} = \frac{1}{2} \quad \text{for a total of} \quad \frac{1}{2} + \frac{1}{2} = 1$$

9.

$$\int_{0}^{1} [0 - (-x)\,dx] = \int_{0}^{1} x\,dx = \frac{1}{2}$$

13.

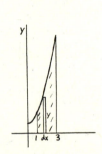

$$\int_{1}^{3} (x^2 + 1)\,dx = \frac{1}{3}x^3 + x\Big|_{1}^{3}$$

$$= (9 + 3) - \left(\frac{1}{3} + 1\right)$$

$$= 12 - \frac{4}{3} = \frac{32}{3}$$

17.

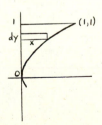

$$\int_{0}^{1} x\,dy = \int_{0}^{1} y^2\,dy = \frac{1}{3}y^3\Big|_{0}^{1} = \frac{1}{3}$$

21.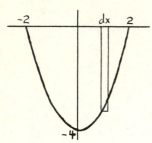

$$\int_{-2}^{2} [0 - (x^2 - 4)]\,dx = \int_{-2}^{2} (-x^2 + 4)\,dx$$

$$= -\tfrac{1}{3}x^3 + 4x\,|_{-2}^{2}$$

$$= \left(-\tfrac{8}{3} + 8\right) - \left(\tfrac{8}{3} - 8\right)$$

$$= -\tfrac{8}{3} + 8 - \tfrac{8}{3} + 8$$

$$= 16 - \tfrac{16}{3} = \tfrac{32}{3}$$

25. To see where the curves intersect, we need to solve the equations simultaneously:

$$y^2 = x - 1$$
$$y = x - 3$$
$$y^2 - y = 2 \text{ (subtracting)}$$
$$y^2 - y - 2 = 0$$
$$(y - 2)(y + 1) = 0$$
$$y = -1,\ 2$$

The points are (2,-1) and (5,2).

Note that in the resulting region the typical element should be drawn sideways. Solving the given equations for x in terms of y, we get

$x = y^2 + 1$ and $x = y + 3$,

respectively. The length of the typical element is now seen to be:

$(y + 3) - (y^2 + 1)$.

Thus

$$\int_{-1}^{2} [(y + 3) - (y^2 + 1)]\,dy = \int_{-1}^{2} (y + 3 - y^2 - 1)\,dy$$

$$= \int_{-1}^{2} (-y^2 + y + 2)\,dy$$

$$= -\tfrac{1}{3}y^3 + \tfrac{1}{2}y^2 + 2y\,|_{-1}^{2}$$

$$= \left(-\tfrac{8}{3} + 2 + 4\right) - \left(\tfrac{1}{3} + \tfrac{1}{2} - 2\right)$$

$$= -\tfrac{8}{3} + 2 + 4 - \tfrac{1}{3} - \tfrac{1}{2} + 2$$

$$= -\tfrac{8}{3} - \tfrac{1}{3} - \tfrac{1}{2} + 8 = \tfrac{9}{2}$$

Section 4.7

1. $$\int_{1}^{\infty} \tfrac{2}{x^4}\,dx = \lim_{b \to \infty} \int_{1}^{b} 2x^{-4}\,dx = \lim_{b \to \infty} \left(2\,\tfrac{x^{-3}}{-3}\right)\Big|_{1}^{b}$$

$$= \lim_{b \to \infty} \left(-\tfrac{2}{3x^3}\right)\Big|_{1}^{b} = \lim_{b \to \infty} \left(-\tfrac{2}{3b^3} + \tfrac{2}{3}\right)$$

$$= \frac{2}{3}$$

5. $\displaystyle\int_0^\infty \frac{x}{(x^2 + 4)^2}dx = \lim_{b\to\infty} \int_0^b (x^2 + 4)^{-2}x\,dx$

$\displaystyle = \lim_{b\to\infty} \frac{1}{2} \int_0^b (x^2 + 4)^{-2}(2x\,dx)$ $\qquad u = x^2 + 4$
$\qquad\qquad\qquad\qquad\qquad\qquad\qquad\qquad du = 2x\,dx$

$\displaystyle = \lim_{b\to\infty} \frac{1}{2} \left.\frac{(x^2 + 4)^{-1}}{-1}\right|_0^b$

$\displaystyle = \lim_{b\to\infty} \left(-\left.\frac{1}{2(x^2 + 4)}\right)\right|_0^b$

$\displaystyle = \lim_{b\to\infty} \left(-\frac{1}{2(b^2 + 4)} + \frac{1}{8}\right) = \frac{1}{8}$

9.

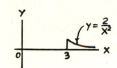

$\displaystyle\int_3^\infty \frac{2}{x^2}dx = \lim_{b\to\infty} \int_3^b 2x^{-2}dx = \lim_{b\to\infty} 2\left.\frac{x^{-1}}{-1}\right|_3^b$

$\displaystyle = \lim_{b\to\infty}\left(-\frac{2}{x}\right)\Big|_3^b = \lim_{b\to\infty}\left(-\frac{2}{b} + \frac{2}{3}\right) = \frac{2}{3}$

13.

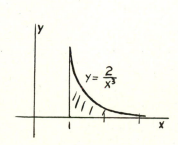

$\displaystyle\int_1^\infty \frac{2}{x^3}dx = \lim_{b\to\infty} \int_1^b 2x^{-3}dx$

$\displaystyle = \lim_{b\to\infty}\left(2\frac{x^{-2}}{-2}\right)\Big|_1^b$

$\displaystyle = \lim_{b\to\infty}\left(-\frac{1}{x^2}\right)\Big|_1^b$

$\displaystyle = \lim_{b\to\infty}\left(-\frac{1}{b^2} + 1\right) = 1$

17.

$$A = \int_{-\infty}^{0} \left[0 - \frac{1}{(2x - 3)^3} \right] dx$$

$$= \lim_{b \to -\infty} \int_{b}^{0} \left[-(2x - 3)^{-3} \right] dx$$

$$= \lim_{b \to -\infty} \frac{1}{2} \int_{b}^{0} \left[-(2x - 3)^{-3} \right] 2 \, dx \qquad\qquad \begin{array}{l} u = 2x - 3 \\ du = 2 \, dx \end{array}$$

$$= \lim_{b \to -\infty} \frac{1}{2} \left(- \left. \frac{(2x - 3)^{-2}}{-2} \right) \right|_{b}^{0}$$

$$= \lim_{b \to -\infty} \frac{1}{4} \left. \frac{1}{(2x - 3)^2} \right|_{b}^{0}$$

$$= \lim_{b \to -\infty} \left(\frac{1}{4} \frac{1}{9} - \frac{1}{4} \frac{1}{(2b - 3)^2} \right) = \frac{1}{36}$$

21.

vertical asymptote: x = 1

We avoid the vertical asymptote by integrating from $1 + \epsilon$ to 5:

$$\int_{1}^{5} \frac{dx}{\sqrt{x - 1}} = \lim_{\epsilon \to 0} \int_{1+\epsilon}^{5} (x - 1)^{-1/2} dx \qquad\qquad u = x - 1; \ du = dx$$

$$= \lim_{\epsilon \to 0} 2(x - 1)^{1/2} \Big|_{1+\epsilon}^{5}$$

$$= \lim_{\epsilon \to 0} \left[2(4)^{1/2} - 2(\epsilon)^{1/2} \right] = 2\sqrt{4} = 4$$

25. $\displaystyle \int_{1}^{\infty} \frac{1}{\sqrt{x}} dx = \lim_{b \to \infty} \int_{1}^{b} x^{-1/2} dx = \lim_{b \to \infty} 2x^{1/2} \Big|_{1}^{b}$

$$= \lim_{b \to \infty} (2\sqrt{b} - 2) = \infty$$

29. $\displaystyle \int_{0}^{4} \frac{x}{(9 - x^2)^2} dx = \int_{0}^{3} \frac{x}{(9 - x^2)^2} dx + \int_{3}^{4} \frac{x}{(9 - x^2)^2} dx$

$$= \lim_{\epsilon \to 0} \int_{0}^{3-\epsilon} (9 - x^2)^{-2} x \, dx + \lim_{\eta \to 0} \int_{3+\eta}^{4} (9 - x^2)^{-2} x \, dx$$

Let $u = 9 - x^2$; then $du = -2x\,dx$:

$$= \lim_{\epsilon \to 0} \left(-\frac{1}{2} \frac{(9 - x^2)^{-1}}{-1} \right)\Bigg|_0^{3-\epsilon} + \lim_{\eta \to 0} \left(-\frac{1}{2} \frac{(9 - x^2)^{-1}}{-1} \right)\Bigg|_{3+\eta}^4$$

$$= \lim_{\epsilon \to 0} \frac{1}{2(9 - x^2)}\Bigg|_0^{3-\epsilon} + \lim_{\eta \to 0} \frac{1}{2(9 - x^2)}\Bigg|_{3+\eta}^4$$

$$= \lim_{\epsilon \to 0} \left\{ \frac{1}{2[9 - (3 - \epsilon)^2]} - \frac{1}{18} \right\} + \lim_{\eta \to 0} \left\{ \frac{1}{2(-7)} - \frac{1}{2[9 - (3 + \eta)^2]} \right\}$$

Neither limit exists.

Section 4.8

1. From $\frac{dy}{dx} = 3x$, we get $y = \frac{3}{2}x^2 + C$. Substituting $(0,1)$ in the equation, we get $1 = 0 + C$ or $C = 1$. The resulting function is

$$y = \frac{3}{2}x^2 + 1$$

5. From $\frac{dy}{dx} = 3x^2 + 2$, we get $y = x^3 + 2x + C$. Substituting $(1,0)$ in the equation, we get $0 = 3 + C$, or $C = -3$. It follows that

$$y = x^3 + 2x - 3$$

9.

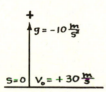

Taking the upward direction as positive (as was done in the examples), we get $g = -10$ m/s^2 and $v_0 = +30$ m/s, since the ball is moving in the positive direction initially. Integrating, we have

$$v = -10t + C$$

If $t = 0$, $v = +30$, so that $30 = 0 + C$. Thus

$$v = -10t + 30$$

To find s, we integrate v:

$$s = -5t^2 + 30t + k$$

If the origin is on the ground, as in the figure, $s = 0$ when $t = 0$. Thus $0 = 0 + 0 + k$ and

$$s = -5t^2 + 30t$$

To find how high the ball rises, we first determine how long it takes to reach the highest point by setting v to 0:

$$v = -10t + 30 = 0, \text{ whence } t = 3$$

Thus

$$s = -5t^2 + 30t|_{t=3} = -45 + 90 = 45$$

In the coordinate system chosen, 45 corresponds to 45 m above the ground.

13.

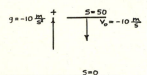

Taking the upward direction as positive, $g = -10 \text{ m/s}^2$, as usual, and $v_0 = -10 \text{ m/s}$ (since the object is moving in the downward direction). Thus

$$v = -10t + C$$

and

$$v = -10t - 10 \qquad \text{if } t = 0, v = -10 \text{ m/s}$$

Hence

$$s = -5t^2 - 10t + k \qquad \text{integrating } v$$

Now observe that when $t = 0$, $s = 50$ m. So

$$50 = 0 + 0 + k$$

and

$$s = -5t^2 - 10t + 50$$

To see how long it takes the object to reach the ground, we let $s = 0$ and solve for t:

$$0 = -5t^2 - 10t + 50$$

$$t^2 + 2t - 10 = 0 \qquad \text{dividing by } -5$$

$$t = \frac{-2 \pm \sqrt{4 + 40}}{2} = \frac{-2 \pm 2\sqrt{11}}{2} = -1 \pm \sqrt{11}$$

Taking the positive root, $t = -1 + \sqrt{11} \approx 2.3$ s. Finally, when $t = 2.3$,

$$v = -10(2.3) - 10 = -33$$

We conclude that the velocity is 33 m/s in the downward direction.

17. Taking the upward direction as positive, $g = -10 \text{ m/s}^2$ and $v_0 = +10 \text{ m/s}$ (since the object is moving in the upward direction). Thus

$$v = -10t + C$$

and

$$v = -10t + 10 \qquad \text{if } t = 0, \ v = +10 \text{ m/s}$$

So

$$s = -5t^2 + 10t + k \qquad \text{integrating } v$$

and

$$s = -5t^2 + 10t + 40 \qquad \text{if } t = 0, \ s = 40 \text{ m}$$

To see how long the object travels before striking the ground, we let $s = 0$:

$$0 = -5t^2 + 10t + 40$$

$$t^2 - 2t - 8 = 0 \qquad \text{dividing by } -5$$

$$(t - 4)(t + 2) = 0$$

$$t = +4 \qquad \text{positive root}$$

From $v = -10t + 10$, we get $v = -30$ m/s, that is, the object is moving in the downward direction.

21. If we let the direction of motion be the positive direction, then $a = -7$ and $v_0 = 28$. For convenience, we let $s = 0$ be the point at which the car starts to decelerate. We now get

$$v = -7t + 28$$

$$s = -\frac{7}{2}t^2 + 28t + 0 \qquad k = 0 \quad (\text{since } s = 0 \text{ when } t = 0)$$

If we let $v = 0$, we find that the car stops in four seconds. From

$$s = -\frac{7}{2}t^2 + 28t \Big|_{t=4} = 56$$

we see that the car stops in 56 m.

25.
$$a = \frac{t}{(t^2 + 1)^2}$$

$$v = \int \frac{t}{(t^2 + 1)^2}\, dt = \int (t^2 + 1)^{-2}t\, dt \qquad u = t^2 + 1; \ du = 2t\, dt$$

$$= \frac{1}{2} \int (t^2 + 1)^{-2}(2t\, dt)$$

$$= \frac{1}{2} \frac{(t^2 + 1)^{-1}}{-1} + C = -\frac{1}{2(t^2 + 1)} + C$$

Since $v = 10$ when $t = 0$, we get from

$$v = -\frac{1}{2(t^2 + 1)} + C$$

$$10 = -\frac{1}{2(0 + 1)} + C \quad \text{or} \quad C = \frac{21}{2}$$

So

$$v = -\frac{1}{2(t^2 + 1)} + \frac{21}{2}$$
$$= \frac{1}{2}\left(21 - \frac{1}{t^2 + 1}\right)$$

29. Given: $i = 0.010t + 0.10$ and $q = 0.030\,C$ when $t = 0$.

$$q = \int i\,dt = \int (0.010t + 0.10)dt = 0.010\left(\frac{t^2}{2}\right) + 0.10t + C$$

Substituting $t = 0$ and $q = 0.030$, we get

$$0.030 = 0 + C$$

so that

$$q = 0.010\left(\frac{t^2}{2}\right) + 0.10t + 0.030\big|_{t=3.0} = 0.38\,C$$

Section 4.9

1. If $n = 6$, we have $h = \frac{b - a}{6} = \frac{1}{3}$. Thus $x_0 = 1$, $x_1 = \frac{4}{3}$, $x_2 = \frac{5}{3}$, $x_3 = \frac{6}{3}$, $x_4 = \frac{7}{3}$, $x_5 = \frac{8}{3}$, $x_6 = \frac{9}{3} = 3$

By the trapezoidal rule:

$$\int_1^3 x^2\,dx \approx \frac{1}{3}\left[\frac{1}{2}(1)^2 + \left(\frac{4}{3}\right)^2 + \left(\frac{5}{3}\right)^2 + \left(\frac{6}{3}\right)^2 + \left(\frac{7}{3}\right)^2 + \left(\frac{8}{3}\right)^2 + \frac{1}{2}\left(\frac{9}{3}\right)^2\right]$$

$$= 8.704$$

By Simpson's rule:

$$\int_1^3 x^2\,dx \approx \frac{1/3}{3}\left[1^2 + 4\left(\frac{4}{3}\right)^2 + 2\left(\frac{5}{3}\right)^2 + 4\left(\frac{6}{3}\right)^2 + 2\left(\frac{7}{3}\right)^2 + 4\left(\frac{8}{3}\right)^2 + \left(\frac{9}{3}\right)^2\right]$$

$$= 8.667$$

By direct integration:

$$\int_1^3 x^2\,dx = \frac{1}{3}x^3\big|_1^3 = \frac{26}{3}$$

5. If $n = 4$, we have $h = \frac{b - a}{4} = \frac{1}{2}$. Thus $x_0 = 0$, $x_1 = \frac{1}{2}$, $x_2 = 1$, $x_3 = \frac{3}{2}$, $x_4 = 2$

By the trapezoidal rule:

$$\int_0^2 \sqrt{1 + x}\,dx \approx \frac{1}{2}\left[\frac{1}{2}\sqrt{1} + \sqrt{\frac{3}{2}} + \sqrt{2} + \sqrt{\frac{5}{2}} + \frac{1}{2}\sqrt{3}\right] = 2.793$$

By Simpson's rule:

$$\int_0^2 \sqrt{1 + x}\,dx \approx \frac{1/2}{3}\left[\sqrt{1} + 4\sqrt{\frac{3}{2}} + 2\sqrt{2} + 4\sqrt{\frac{5}{2}} + \sqrt{3}\right]$$

$$= \frac{1}{6}\left[1 + 4\sqrt{1.5} + 2\sqrt{2} + 4\sqrt{2.5} + \sqrt{3}\right]$$

The sequence is

1 ⊞ 4 ⊠ 1.5 🔽 ⊞ 2 ⊠ 2 🔽 ⊞ 4 ⊠ 2.5 🔽 ⊞ 3 🔽

═ ÷ 6 ═

The answer is 2.797

9. $\displaystyle\int_{-1}^{2} \frac{dx}{x^3 + 2}$ $h = \dfrac{2 - (-1)}{12} = 0.25$

$x_0 = -1$, $x_1 = -1 + 0.25 = -0.75$, $x_2 = -0.75 + 0.25 = -0.5$,

$x_3 = -0.5 + 0.25 = -0.25$, etc.

The function values are listed next:

$f(x_0) = 1/(-1 + 2) = 1$

$f(x_1) = 1/[(-0.75)^3 + 2] = 0.63366$

$f(x_2) = 1/[(-0.5)^3 + 2] = 0.53333$

$f(x_3) = 1/[(-0.25)^3 + 2] = 0.50394$

$f(x_4) = 1/(0 + 2) = 0.5$

$f(x_5) = 1/(0.25^3 + 2) = 0.49612$

$f(x_6) = 1/(0.5^3 + 2) = 0.47059$

$f(x_7) = 1/(0.75^3 + 2) = 0.41290$

$f(x_8) = 1/(1 + 2) = 0.33333$

$f(x_9) = 1/(1.25^3 + 2) = 0.25296$

$f(x_{10}) = 1/(1.5^3 + 2) = 0.18605$

$f(x_{11}) = 1/(1.75^3 + 2) = 0.13588$

$f(x_{12}) = 1/(2^3 + 2) = 0.1$

By the trapezoidal rule:

$$\int_{-1}^{2} \frac{dx}{x^3 + 2} \approx 0.25\left[\tfrac{1}{2}(1) + 0.63366 + 0.53333 + \ldots + \tfrac{1}{2}(0.1)\right] = 1.252$$

13. Since $n = 6$, we get $h = \dfrac{b - a}{6} = \dfrac{1}{2}$. Thus $x_0 = 1$, $x_1 = \dfrac{3}{2}$, $x_2 = 2$,

$x_3 = \dfrac{5}{2}$, $x_4 = 3$, $x_5 = \dfrac{7}{2}$, $x_6 = 4$

By Simpson's rule:

$$\int_{1}^{4} \sqrt{1 + x^2}\,dx \approx \frac{1/2}{3}\Big[\sqrt{1 + 1^2} + 4\sqrt{1 + (3/2)^2} + 2\sqrt{1 + 2^2}$$
$$+ 4\sqrt{1 + (5/2)^2} + 2\sqrt{1 + 3^2} + 4\sqrt{1 + (7/2)^2}$$
$$+ \sqrt{1 + 4^2}\Big] = 8.146$$

17. From the table, $h = 1$. Using the given y-values, we get, by Simpson's rule:

$$\tfrac{1}{3}[1.3 + 4(1.9) + 2(3.2) + 4(3.8) + 2(4.7) + 4(6.8) + 2(10.2)$$
$$+ 4(15.6) + 20.3] = 56.7$$

1. Subdivide the interval $[0,3]$ into n equal parts, each of length $\frac{3}{n}$. Thus $\Delta x_i = \frac{3}{n}$.

   Choosing the right endpoint of each subinterval for x, we get
   $$x_1 = 1 \cdot \frac{3}{n}, \; x_2 = 2 \cdot \frac{3}{n}, \; \ldots, \; x_i = i \cdot \frac{3}{n}, \; \ldots, \; x_n = n \cdot \frac{3}{n} = 3$$
   For the altitudes, we get from $f(x) = 3x^2$
   $$f(x_1) = 3(1)^2 \frac{9}{n^2}, \; f(x_2) = 3(2)^2 \frac{9}{n^2}, \; \ldots, \; f(x_i) = 3(i)^2 \frac{9}{n^2},$$
   $$\ldots, \; f(x_n) = 3(n)^2 \frac{9}{n^2}$$

   Since $\Delta x_i = \frac{3}{n}$, the sum of the areas is
   $$\sum_{i=1}^{n} f(x_i) \Delta x_i = \sum_{i=1}^{n} 3i^2 \frac{9}{n^2} \frac{3}{n} = \frac{81}{n^3} \sum_{i=1}^{n} i^2 = \frac{81}{n^3} \frac{n(n+1)(2n+1)}{6}$$
   $$= \frac{27n(n+1)(2n+1)}{2n^3}$$

   by Formula B. Hence
   $$\int_0^3 3x^2 dx = \lim_{n \to \infty} \frac{27n(n+1)(2n+1)}{2n^3} = \lim_{n \to \infty} \frac{54n^3 + 81n^2 + 27n}{2n^3}$$
   $$= \lim_{n \to \infty} \frac{54 + 81/n + 27/n^2}{2} = 27$$

5. $\int (1 - x^2)^5 x \, dx$ Let $u = 1 - x^2$; then $du = -2x \, dx$

   $$\int (1 - x^2)^5 x \, dx = -\frac{1}{2} \int (1 - x^2)^5 (-2x) \, dx = -\frac{1}{2} \int u^5 \, du$$

   $$= -\frac{1}{2} \frac{u^6}{6} + C = -\frac{1}{12}(1 - x^2)^6 + C$$

9. $\int (x^3 + 1)^2 (3x) \, dx$.   If $u = x^3 + 1$, then $du = 3x^2 \, dx$, which is different

   from $3x \, dx$. Since the substitution cannot be made, we multiply out the integrand and integrate term by term:

   $$\int (x^6 + 2x^3 + 1)(3x) \, dx = \int (3x^7 + 6x^4 + 3x) \, dx$$
   $$= \frac{3}{8}x^8 + \frac{6}{5}x^5 + \frac{3}{2}x^2 + C$$

13. $\int (x - 2)\sqrt{x^2 - 4x} \, dx = \int (x^2 - 4x)^{1/2}(x - 2) \, dx$

   Let $u = x^2 - 4x$; then $du = (2x - 4) \, dx = 2(x - 2) \, dx$

   $$\int (x^2 - 4x)^{1/2}(x - 2) \, dx = \frac{1}{2} \int (x^2 - 4x)^{1/2} 2(x - 2) \, dx$$

$$= \frac{1}{2} \int u^{1/2} du = \frac{1}{2} \frac{u^{3/2}}{3/2} + C$$

$$= \frac{1}{3}(x^2 - 4x)^{3/2} + C$$

17.

$$\int_{-2}^{2} x \, dy = \int_{-2}^{2} (4 - y^2) \, dy$$

$$= 4y - \frac{y^3}{3} \Big|_{-2}^{2}$$

$$= \left(8 - \frac{8}{3}\right) - \left(-8 + \frac{8}{3}\right)$$

$$= 16 - \frac{16}{3} = \frac{32}{3}$$

21.
$$\int_{-\infty}^{0} \frac{2x}{(x^2 + 4)^2} dx = \lim_{b \to -\infty} \int_{b}^{0} (x^2 + 4)^{-2} (2x) \, dx$$

$$= \lim_{b \to -\infty} \frac{(x^2 + 4)^{-1}}{-1} \Big|_{b}^{0} \qquad\qquad u = x^2 + 4$$
$$\qquad\qquad\qquad du = 2x \, dx$$

$$= \lim_{b \to -\infty} \left(- \frac{1}{x^2 + 4}\right) \Big|_{b}^{0}$$

$$= \lim_{b \to -\infty} \left(-\frac{1}{4} + \frac{1}{b^2 + 4}\right) = -\frac{1}{4}$$

25.
$$\int_{2}^{\infty} \frac{1}{x^3} dx = \lim_{b \to \infty} \int_{2}^{b} x^{-3} dx = \lim_{b \to \infty} \frac{x^{-2}}{-2} \Big|_{2}^{b} = \lim_{b \to \infty} \left(-\frac{1}{2x^2}\right) \Big|_{2}^{b}$$

$$= \lim_{b \to \infty} \left(-\frac{1}{2b^2} + \frac{1}{8}\right) = \frac{1}{8}$$

29. $q = \int i \, dt = \int 3.08 t^{1/2} dt = 3.08 \left(\frac{2}{3}\right) t^{3/2} + C.$ If $t = 0$, $q = 0$, so that
$C = 0.$ Thus

$$q = 3.08\left(\frac{2}{3}\right) t^{3/2} \Big|_{t=1.75} = 4.75 \, C$$

33. Since $n = 6$, $\frac{b - a}{6} = \frac{3}{6} = \frac{1}{2}$. Thus $x_0 = -1$, $x_1 = -\frac{1}{2}$, $x_2 = 0$,
$x_3 = \frac{1}{2}$, $x_4 = 1$, $x_5 = \frac{3}{2}$, $x_6 = 2$

By Simpson's rule:
$$\int_{-1}^{2} \frac{dx}{x + 3} \approx \frac{1/2}{3}\left[\frac{1}{2} + 4\left(\frac{1}{5/2}\right) + 2\left(\frac{1}{3}\right) + 4\left(\frac{1}{7/2}\right) + 2\left(\frac{1}{4}\right)\right.$$
$$\left. + 4\left(\frac{1}{9/2}\right) + \frac{1}{5}\right] = 0.916$$

CHAPTER 5   APPLICATIONS OF THE INTEGRAL

<u>Section 5.1</u>

1. $f_{av} = \frac{1}{16 - 1} \int_1^{16} x^{1/2}dx = \frac{1}{15} \frac{2}{3}x^{3/2}\Big|_1^{16} = \frac{2}{45}(16^{3/2} - 1)$

$= \frac{2}{45}(64 - 1) = \frac{2(63)}{45} = \frac{14}{5}$

5. $f_{rms}^2 = \frac{1}{2 - 1} \int_1^2 \left(\frac{1}{x}\right)^2 dx = \int_1^2 x^{-2}dx = \frac{x^{-1}}{-1}\Big|_1^2 = -\frac{1}{x}\Big|_1^2 = -\frac{1}{2} + 1 = \frac{1}{2}$

Thus $f_{rms} = \sqrt{\frac{1}{2}} = \frac{\sqrt{2}}{2}$.

9.   Taking the upward direction as positive, we have $g = -10$ m/s^2.   Also,
when $t = 0$, we have $v = 0$ and $s = 180$.   Thus

$v = -10t$ since $v_0 = 0$

$s = -5t^2 + k$

$s = -5t^2 + 180$ since $s = 180$ when $t = 0$

Now let $s = 0$ and solve for t:

$0 = -5t^2 + 180$ or $t = 6$ s

So it takes the object 6 s to reach the ground.   We now get

$v_{av} = \frac{1}{6 - 0} \int_0^6 (-10t)dt = -\frac{1}{6}(5t^2)\Big|_0^6 = -30$ m/s

so the average velocity is 30 m/s in the downward direction.

13. $i_{rms}^2 = \frac{1}{3-0} \int_0^3 (1 - t^2)^2 dt$ (leaving out final zeros)

$= \frac{1}{3} \int_0^3 (1 - 2t^2 + t^4)dt = \frac{1}{3}\left(t - \frac{2}{3}t^3 + \frac{1}{5}t^5\right)\Big|_0^3$

$= \frac{1}{3}\left(3 - 18 + \frac{243}{5}\right) = \frac{1}{3} \frac{-75 + 243}{5} = \frac{168}{15}$

Since $R = 5$, we get

$P = i_{rms}^2 R = \frac{168}{15}(5) = 56$ W to two significant digits.

1.

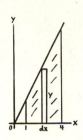

Volume of disk: $\pi y^2 dx$

$$V = \int_1^4 \pi(2x)^2 dx = \pi \int_1^4 4x^2 dx$$

$$= \pi\left(\tfrac{4}{3}\right)x^3\big|_1^4 = \tfrac{4\pi}{3}(64 - 1)$$

$$= \tfrac{252\pi}{3} = 84\pi$$

5.

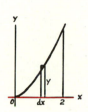

$$V = \pi \int_0^2 (x^{3/2})^2 dx$$

$$= \pi \int_0^2 x^3 dx$$

$$= \pi\left(\tfrac{1}{4}x^4\right)\Big|_0^2 = 4\pi$$

9. $V = \int_1^3 \pi\left(\sqrt{x^2 + 1}\right)^2 dx = \pi \int_1^3 (x^2 + 1) dx = \tfrac{32\pi}{3}$

13.

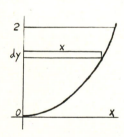

Volume of disk: $\pi x^2 dy$

From $y = \tfrac{1}{2}x^2$, we get $x^2 = 2y$

$$V = \pi \int_0^2 2y \, dy = \pi y^2\big|_0^2 = 4\pi$$

17.

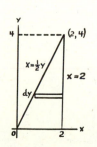

Note that the typical element has to be drawn horizontally to generate washers.

Volume of typical washer: $\pi\left(x_2^2 - x_1^2\right)dy$

$$= \pi\left[2^2 - \left(\tfrac{1}{2}y\right)^2\right]$$

Integrating from $y = 0$ to $y = 4$, we get

$$V = \int_0^4 \pi\left[2^2 - \left(\tfrac{1}{2}y\right)^2\right]dy = \pi \int_0^4\left(4 - \tfrac{1}{4}y^2\right)dy$$

$$= \pi\left(4y - \tfrac{1}{12}y^3\right)\Big|_0^4 = \pi\left(4^2 - \tfrac{1}{12}4^3\right)$$

$$= \pi\left(4^2 - \tfrac{1}{3} \cdot 4^2\right) = 4^2\pi\left(1 - \tfrac{1}{3}\right) = 16\pi \cdot \tfrac{2}{3} = \tfrac{32\pi}{3}$$

21. Since $y = \tfrac{1}{2}x$, we have $x = 2y$

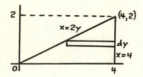

Volume of typical washer: $\pi\left(x_2^2 - x_1^2\right)dy$

$$= \pi\left[4^2 - (2y)^2\right]dy$$

Integrating from $y = 0$ to $y = 2$, we get

$$V = \int_0^2 \pi\left[4^2 - (2y)^2\right]dy = \pi \int_0^2\left(16 - 4y^2\right)dy$$

$$= \pi\left(16y - \tfrac{4}{3}y^3\right)\Big|_0^2 = \pi\left(16 \cdot 2 - \tfrac{4}{3} \cdot 2^3\right)$$

$$= \pi\left(32 - \tfrac{1}{3} \cdot 32\right) = 32\pi\left(1 - \tfrac{1}{3}\right)$$

$$= 32\pi\left(\tfrac{2}{3}\right) = \tfrac{64\pi}{3}$$

25. To find the points of intersection, we need to solve the equations simultaneously:

$$x = y^2$$
$$\underline{x = y + 2}$$
$$0 = y^2 - y - 2 \quad \text{(subtracting)}$$

$$(y + 1)(y - 2) = 0$$
$$y = -1, \; 2$$

The points of intersection are $(1, -1)$ and $(4, 2)$.

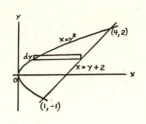

Note that the typical element has to be drawn horizontally. Now recall that the volume of a typical washer is

$$\pi(x_2^2 - x_1^2)\,dy$$

$$= \pi[(y + 2)^2 - (y^2)^2]\,dy$$

Integrating from $y = -1$ to $y = 2$, we get

$$V = \int_{-1}^{2} \pi[(y + 2)^2 - (y^2)^2]\,dy = \pi \int_{-1}^{2} (y^2 + 4y + 4 - y^4)\,dy$$

$$= \pi\left(\tfrac{1}{3}y^3 + 2y^2 + 4y - \tfrac{1}{5}y^5\right)\Big|_{-1}^{2}$$

$$= \pi\left[\left(\tfrac{8}{3} + 8 + 8 - \tfrac{32}{5}\right) - \left(-\tfrac{1}{3} + 2 - 4 + \tfrac{1}{5}\right)\right]$$

$$= \pi\left(\tfrac{8}{3} + 8 + 8 - \tfrac{32}{5} + \tfrac{1}{3} - 2 + 4 - \tfrac{1}{5}\right) = \pi\left(\tfrac{9}{3} - \tfrac{33}{5} + 18\right)$$

$$= \pi\left(21 - \tfrac{33}{5}\right) = \pi\,\frac{105 - 33}{5} = \frac{72\pi}{5}$$

29. By turning the tank on the side, we can find the volume of the water by rotating the region in the figure about the x-axis. (Since the radius is 12 ft, note that the equation of the circle is $x^2 + y^2 = 12^2$.)

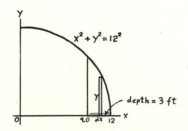

From $x^2 + y^2 = 144$, we get $y^2 = 144 - x^2$

Volume of typical disk: $\pi y^2\,dx = \pi(144 - x^2)\,dx$

Integrating from $x = 9$ to $x = 12$,

$$V = \pi \int_{9}^{12} (144 - x^2)\,dx = \pi\left(144x - \tfrac{1}{3}x^3\right)\Big|_{9}^{12}$$

$$= \pi\left[\left(144 \cdot 12 - \tfrac{1}{3}(12)^3\right) - \left(144 \cdot 9 - \tfrac{1}{3} \cdot 9^3\right)\right]$$

$$= \pi[(1728 - 576) - (1296 - 243)] = 99\pi \approx 310 \text{ ft}^3$$

33. $y = \dfrac{1}{x^{3/4}} = x^{-3/4}$

Volume of typical disk: $\pi y^2 dx = \pi\left(x^{-3/4}\right)^2 dx = \pi x^{-3/2} dx$

$$V = \pi \int_4^\infty x^{-3/2}dx = \lim_{b \to \infty} \pi \int_4^b x^{-3/2}dx = \lim_{b \to \infty} \pi(-2)x^{-1/2}\Big|_4^b$$

$$= \lim_{b \to \infty} \frac{-2\pi}{\sqrt{x}}\Big|_4^b = \lim_{b \to \infty}\left[\frac{-2\pi}{\sqrt{b}} + \frac{2\pi}{\sqrt{4}}\right] = \frac{2\pi}{2} = \pi$$

$$A = \int_4^\infty x^{-3/4}dx = \lim_{b \to \infty} \int_4^b x^{-3/4}dx = \lim_{b \to \infty} 4x^{1/4}\Big|_4^b$$

$$= \lim_{b \to \infty}\left(4b^{1/4} - 4(4)^{1/4}\right) = \infty \quad \text{(does not exist)}$$

Section 5.3

1. Volume of shell: 2π (radius) $\cdot$ (height) $\cdot$ (thickness)

$$= 2\pi \cdot x \cdot y \cdot dx = 2\pi x \cdot x\,dx$$

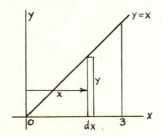

$$V = \int_0^3 2\pi x(x)\,dx = 2\pi \int_0^3 x^2 dx = 2\pi \left.\frac{x^3}{3}\right|_0^3 = 18\pi$$

5. Substituting $y = x^2$ in $y^2 = 8x$, we get

$$x^4 = 8x$$
$$x^4 - 8x = 0$$
$$x(x^3 - 8) = 0$$
$$x = 0, 2$$

The points of intersection are $(0,0)$ and $(2,4)$.

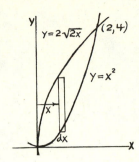

The functions are

$$y = x^2$$

and

$$y = \sqrt{8x} = 2\sqrt{2x}$$

Note that the height of the typical element is

$$(2\sqrt{2x} - x^2)\,dx$$

Volume of shell: 2π (radius) $\cdot$ (height) $\cdot$ (thickness)

$$= 2\pi \cdot x \cdot (2\sqrt{2x} - x^2)\,dx$$

$$V = \int_0^2 2\pi x(2\sqrt{2x} - x^2)\,dx = 2\pi \int_0^2 (2\sqrt{2}\,x^{3/2} - x^3)\,dx$$

$$= 2\pi\left[2\sqrt{2}\left(\tfrac{2}{5}\right)x^{5/2} - \tfrac{1}{4}x^4\right]_0^2 = 2\pi\left[2\sqrt{2}\left(\tfrac{2}{5}\right)2^{5/2} - 4\right]$$

$$= 2\pi\left[\tfrac{4}{5}(2)^{6/2} - 4\right] = 2\pi\left(\tfrac{4}{5} \cdot 8 - 4\right) = 2\pi\,\frac{32 - 20}{5} = \frac{24\pi}{5}$$

9.

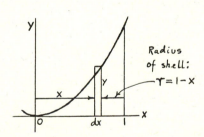

Radius of shell: $r = 1 - x$

Volume of typical shell: 2π (radius) $\cdot$ (height) $\cdot$ (thickness)

$$= 2\pi(1 - x) \cdot x^2\,dx$$

$$V = \int_0^1 2\pi(1 - x)x^2\,dx = 2\pi \int_0^1 (x^2 - x^3)\,dx$$

$$= 2\pi\left(\tfrac{1}{3}x^3 - \tfrac{1}{4}x^4\right)\Big|_0^1 = 2\pi\left(\tfrac{1}{3} - \tfrac{1}{4}\right) = \frac{2\pi}{12} = \frac{\pi}{6}$$

13.

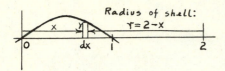

Note that the intercepts of $y = x - x^2 = x(1 - x)$ are $x = 0$ and $x = 1$.

Volume of shell: 2π (radius) $\cdot$ (height) $\cdot$ (thickness)
$$= 2\pi(2 - x) \cdot (x - x^2)dx$$

$$V = \int_0^1 2\pi(2 - x)(x - x^2)dx = 2\pi \int_0^1 (2x - 3x^2 + x^3)dx$$

$$= 2\pi\left(x^2 - x^3 + \tfrac{1}{4}x^4\right)\Big|_0^1 = 2\pi\left(1 - 1 + \tfrac{1}{4}\right) = \tfrac{\pi}{2}$$

17.

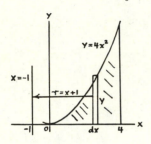

radius of shell: $x - (-1) = x + 1$

Volume of shell: 2π (radius) $\cdot$ (height) $\cdot$ (thickness)
$$= 2\pi(x + 1) \cdot 4x^2dx$$

$$V = \int_0^4 2\pi(x + 1)(4x^2)dx = 8\pi \int_0^4 (x^3 + x^2)dx$$

$$= 8\pi\left(\tfrac{1}{4}x^4 + \tfrac{1}{3}x^3\right)\Big|_0^4 = 8\pi\left(\tfrac{1}{4} \cdot 4^4 + \tfrac{1}{3} \cdot 4^3\right)$$

$$= 8\pi\left(4^3 + \tfrac{1}{3} \cdot 4^3\right) = 8\pi(4^3)\left(1 + \tfrac{1}{3}\right) = 8\pi(64)\left(\tfrac{4}{3}\right)$$

$$= \tfrac{2048\pi}{3}$$

21.

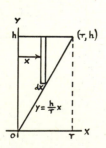

We can obtain a cone by rotating the region in the figure about the y-axis. Note that the slope of the line is $m = \frac{h}{r}$, so that the equation is

$$y = \frac{h}{r}x$$

Height of typical element: $h - \frac{h}{r}x$

Volume of shell: 2π (radius) $\cdot$ (height) $\cdot$ (thickness)

$$= 2\pi \cdot x \cdot \left(h - \frac{h}{r}x\right)dx$$

$$V = \int_0^r 2\pi x\left(h - \frac{h}{r}x\right)dx = 2\pi \int_0^r x \cdot h\left(1 - \frac{1}{r}x\right)dx$$

$$= 2\pi h \int_0^r \left(x - \frac{1}{r}x^2\right)dx = 2\pi h\left(\frac{1}{2}x^2 - \frac{1}{r} \cdot \frac{1}{3}x^3\right)\Big|_0^r$$

$$= 2\pi h\left(\frac{1}{2}r^2 - \frac{1}{3}r^2\right) = 2\pi h r^2\left(\frac{1}{2} - \frac{1}{3}\right)$$

$$= 2\pi h r^2 \frac{1}{6} = \frac{1}{3}\pi r^2 h$$

25. (a)

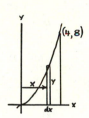

The simplest way to find this volume is by the disk method:

$$\pi y^2 dx$$

Since $y^2 = x^3$, we get

$$V = \pi \int_0^4 x^3 dx = \pi \left.\frac{x^4}{4}\right|_0^4 = 64\pi$$

(b) The simplest way to find this volume is by the shell method. Since $y = x^{3/2}$, we have

$$V = 2\pi \int_0^4 x \cdot x^{3/2}dx = 2\pi \int_0^4 x^{5/2}dx = 2\pi\left(\frac{2}{7}\right)x^{7/2}\Big|_0^4$$

$$= \frac{4\pi}{7}(4)^{7/2} = \frac{4\pi}{7}(128) = \frac{512\pi}{7}$$

(b) alternate

By the washer method, we get, from the typical washer,

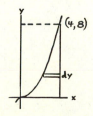

$$\pi(x_2^2 - x_1^2)dy \quad \text{and} \quad x = y^{2/3}$$

$$V = \pi \int_0^8 [4^2 - (y^{2/3})^2]dy,$$

which is harder to evaluate.

(c)

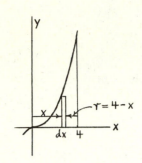

$r = 4 - x$; by shells:

$$V = 2\pi \int_0^4 (4 - x)x^{3/2}dx$$

$$= 2\pi \int_0^4 (4x^{3/2} - x^{5/2})dx$$

$$= 2\pi\left[4\left(\tfrac{2}{5}\right)x^{5/2} - \tfrac{2}{7}x^{7/2}\right]_0^4$$

$$= 2\pi\left[\tfrac{8}{5}(4)^{5/2} - \tfrac{2}{7}(4)^{7/2}\right]$$

$$= 2\pi\left[\tfrac{8}{5}(32) - \tfrac{2}{7}(128)\right]$$

$$= 2\pi(256)\left(\tfrac{1}{5} - \tfrac{1}{7}\right)$$

$$= 512\pi\left(\tfrac{2}{35}\right) = \tfrac{1024\pi}{35}$$

(d)

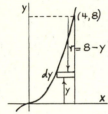

Volume of shell:

$2\pi(\text{radius}) \cdot (\text{height}) \cdot (\text{thickness})$

Since $r = 8 - y$ and the height of the typical element is

$(4 - y^{2/3})dy$,

we get:

$$V = 2\pi \int_0^8 (8 - y)(4 - y^{2/3})dy = 2\pi \int_0^8 (32 - 4y - 8y^{2/3} + y^{5/3})dy$$

$$= 2\pi\left[32y - 2y^2 - 8\left(\tfrac{3}{5}\right)y^{5/3} + \left(\tfrac{3}{8}\right)y^{8/3}\right]_0^8$$

$$= 2\pi\left[256 - 128 - \tfrac{24}{5}(8)^{5/3} + \tfrac{3}{8}(8)^{8/3}\right]$$

$$= 2\pi\left[128 - \tfrac{24}{5}(32) + \tfrac{3}{8}(256)\right]$$

$$= 2\pi\left(128 - \tfrac{768}{5} + \tfrac{768}{8}\right) = 2\pi(128)\left(1 - \tfrac{6}{5} + \tfrac{6}{8}\right)$$

$$= 256\pi\,\frac{20 - 24 + 15}{20} = 256\pi\left(\tfrac{11}{20}\right) = \frac{64\pi(11)}{5} = \frac{704\pi}{5}$$

29.

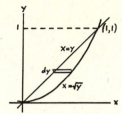

Volume of typical washer: $\pi[(\sqrt{y})^2 - y^2]\,dy$

$$V = \int_0^1 \pi[(\sqrt{y})^2 - y^2]\,dy = \pi \int_0^1 (y - y^2)\,dy$$

$$= \pi\left(\tfrac{1}{2}y^2 - \tfrac{1}{3}y^3\right)\Big|_0^1 = \pi\left(\tfrac{1}{2} - \tfrac{1}{3}\right) = \tfrac{\pi}{6}$$

Section 5.4

1.

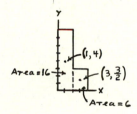

$$\bar{x} = \frac{1\cdot16 + 3\cdot6}{16 + 6} = \frac{34}{22} = \frac{17}{11}$$

$$\bar{y} = \frac{4\cdot16 + \tfrac{3}{2}\cdot6}{16 + 6} = \frac{73}{22}$$

5.

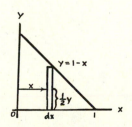

$\bar{x}$: moment of typical element: $xy\,dx$

$$\bar{x} = \frac{M_y}{A} = \frac{\int_0^1 xy\,dx}{\int_0^1 y\,dx} = \frac{\int_0^1 x(1-x)\,dx}{\int_0^1 (1-x)\,dx} = \frac{\frac{1}{2}x^2 - \frac{1}{3}x^3\big|_0^1}{x - \frac{1}{2}x^2\big|_0^1}$$

$$= \frac{\frac{1}{2} - \frac{1}{3}}{1 - \frac{1}{2}} = \frac{\frac{1}{6}}{\frac{1}{2}} = \frac{1}{3}$$

$\bar{y}$: moment of typical element: $\left(\frac{1}{2}y\right)(y\,dx)$

$$\bar{y} = \frac{M_x}{A} = \frac{\int_0^1 \left(\frac{1}{2}y\right)(y\,dx)}{A} = \frac{\frac{1}{2}\int_0^1 (1-x)(1-x)\,dx}{1/2} = 2 \cdot \frac{1}{2} \int_0^1 (1 - 2x + x^2)\,dx$$

$$= x - x^2 + \frac{1}{3}x^3\big|_0^1 = \frac{1}{3}$$

$\bar{y}$ (alternate):

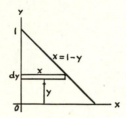

Interchanging the roles of x and y

Moment of typical element with respect to x-axis: $y \cdot x\,dy$

$$\bar{y} = \frac{M_x}{A} = \frac{\int_0^1 yx\,dy}{A} = \frac{\int_0^1 y(1-y)\,dy}{1/2} = \frac{1}{3}$$

9.

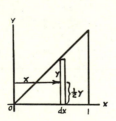

$\bar{x}$: moment of typical element: $xy\,dx$

$$\bar{x} = \frac{M_y}{A} = \frac{\int_0^1 xy\,dx}{\int_0^1 y\,dx} = \frac{\int_0^1 x(x)\,dx}{\int_0^1 x\,dx} = \frac{\frac{1}{3}x^3\big|_0^1}{\frac{1}{2}x^2\big|_0^1} = \frac{\frac{1}{3}}{\frac{1}{2}}$$

$$= \frac{2}{3}$$

$\bar{y}$: moment of typical element: $\left(\frac{1}{2}y\right)(y\,dx)$

$$\bar{y} = \frac{M_x}{A} = \frac{\int_0^1\left(\frac{1}{2}y\right)(y\,dx)}{A} = \frac{\frac{1}{2}\int_0^1 x \cdot x\,dx}{1/2} = \int_0^1 x^2\,dx$$

$$= \frac{1}{3}$$

$\bar{y}$ (alternate): moment (with respect to x-axis) of typical element: $y \cdot$ (height) $\cdot$ dy $= y(1 - y)\,dy$.

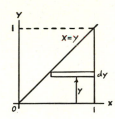

$$\bar{y} = \frac{\int_0^1 y(1 - y)\,dy}{1/2}$$

$$= 2\int_0^1 (y - y^2)\,dy$$

$$= \frac{1}{3}$$

13. Intercepts: $4 - x^2 = 0$ when $x = \pm 2$

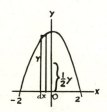

$$A = \int_{-2}^2 (4 - x^2)\,dx = \frac{32}{3}$$

$\bar{y}$: moment of typical element: $\left(\frac{1}{2}y\right)(y\,dx)$

$$\bar{y} = \frac{M_x}{A} = \frac{\int_{-2}^2\left(\frac{1}{2}y\right)(y\,dx)}{A} = \frac{\frac{1}{2}\int_{-2}^2 (4 - x^2)(4 - x^2)\,dx}{32/3}$$

$$= \frac{3}{32} \cdot \frac{1}{2}\int_{-2}^2 (16 - 8x^2 + x^4)\,dx = \frac{3}{64}\left(16x - \frac{8x^3}{3} + \frac{x^5}{5}\right)\Big|_{-2}^2$$

$$= \frac{3}{64}\left(32 - \frac{64}{3} + \frac{32}{5}\right) - \frac{3}{64}\left(-32 + \frac{64}{3} - \frac{32}{5}\right)$$

$$\bar{y} = 2 \cdot \frac{3}{64}\left(32 - \frac{64}{3} + \frac{32}{5}\right) = \frac{3}{32}\frac{480 - 320 + 96}{15}$$

$$= \frac{3}{32}\frac{256}{15} = \frac{8}{5}$$

$\bar{x} = 0$ by symmetry

17.

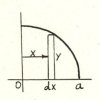

$$y = \frac{1}{\sqrt[3]{x}} = x^{-1/3}$$

$$\bar{x} = \frac{M_y}{A} = \frac{\int_1^8 x \cdot x^{-1/3}dx}{\int_1^8 x^{-1/3}dx} = \frac{\int_1^8 x^{2/3}dx}{\int_1^8 x^{-1/3}dx}$$

$$= \frac{\left(\frac{3}{5}\right)x^{5/3}\Big|_1^8}{\left(\frac{3}{2}\right)x^{2/3}\Big|_1^8} = \frac{\left(\frac{3}{5}\right)(8^{5/3} - 1)}{\left(\frac{3}{2}\right)(8^{2/3} - 1)}$$

$$= \frac{(3/5)(32 - 1)}{(3/2)(4 - 1)} = \frac{(3/5)(31)}{(3/2)(3)} = \frac{93/5}{9/2} = \frac{93}{5}\frac{2}{9} = \frac{31}{5}\frac{2}{3} = \frac{62}{15}$$

$$\bar{y} = \frac{M_x}{A} = \frac{\int_1^8 \left(\frac{1}{2}y\right)y\,dx}{9/2} = \frac{2}{9}\int_1^8 \frac{1}{2}y^2 dx = \frac{1}{9}\int_1^8 (x^{-1/3})^2 dx$$

$$= \frac{1}{9}\int_1^8 x^{-2/3}dx = \frac{1}{9}(3x^{1/3})\Big|_1^8 = \frac{1}{3}(2 - 1) = \frac{1}{3}$$

21.

Since the region is a quarter of a circle, the area is:

$$\frac{1}{4}\pi a^2$$

$$\bar{x} = \frac{M_y}{A} = \frac{\int_0^a x\sqrt{a^2 - x^2}dx}{\frac{1}{4}\pi a^2} = \frac{4}{\pi a^2}\int_0^a (a^2 - x^2)^{1/2}x\,dx$$

Letting $u = a - x^2$; then $du = -2x\,dx$. We now get

$$\frac{4}{\pi a^2}\left(-\frac{1}{2}\right)\int_0^a (a^2 - x^2)^{1/2}(-2x)\,dx = -\frac{2}{\pi a^2}\frac{(a^2 - x^2)^{3/2}\Big|_0^a}{3/2}$$

$$= -\frac{2}{\pi a^2}\left(\frac{2}{3}\right)(a^2 - x^2)^{3/2}\Big|_0^a$$

$$= 0 + \frac{4}{3\pi a^2}(a^2)^{3/2}$$

$$= \frac{4}{3\pi a^2}(a^3) = \frac{4a}{3\pi}$$

By symmetry, $\bar{y} = \frac{4a}{3\pi}$.

25. Substituting $y = 2x$ in $y^2 = 4x$, we obtain:

$$4x^2 = 4x$$
$$x^2 - x = 0$$
$$x(x - 1) = 0$$
$$x = 0, 1$$

The points of intersection are $(0,0)$ and $(1,2)$.

$\bar{x}$: moment (with respect to y-axis) of typical element: $x \cdot$ (height) $\cdot$ dx
$= x(2\sqrt{x} - 2x)dx$

$$\bar{x} = \frac{M_y}{A} = \frac{\displaystyle\int_0^1 x(2\sqrt{x} - 2x)dx}{\displaystyle\int_0^1 (2\sqrt{x} - 2x)dx}$$

$$= \frac{\displaystyle\int_0^1 (2x^{3/2} - 2x^2)dx}{\displaystyle\int_0^1 (2x^{1/2} - 2x)dx}$$

$$= \frac{\frac{4}{5}x^{5/2} - \frac{2}{3}x^3 \Big|_0^1}{\frac{4}{3}x^{3/2} - x^2 \Big|_0^1} = \frac{\frac{4}{5} - \frac{2}{3}}{\frac{4}{3} - 1}$$

$$= \frac{\frac{12 - 10}{15}}{\frac{1}{3}} = \frac{2}{15} \cdot \frac{3}{1} = \frac{2}{5}$$

To obtain $\bar{y}$, note that

(a) distance from x-axis to center of element $= \frac{1}{2}(2\sqrt{x} + 2x)$
(b) height of typical element $= (2\sqrt{x} - 2x)dx$

$$\bar{y} = \frac{M_x}{A} = \frac{\displaystyle\int_0^1 \frac{1}{2}(2\sqrt{x} + 2x)(2\sqrt{x} - 2x)dx}{1/3} = \frac{3}{2}\int_0^1 (4x - 4x^2)dx$$

$$= \frac{3}{2}\left(2x^2 - \frac{4}{3}x^3\right)\Big|_0^1 = \frac{3}{2}\left(2 - \frac{4}{3}\right) = \frac{3}{2} \cdot \frac{2}{3} = 1$$

101

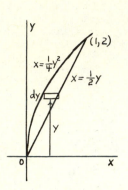

$\bar{y}$ (alternate): moment (with respect to x-axis) of typical element:

$$y \cdot (\text{height}) \cdot dy = y\left(\tfrac{1}{2}y - \tfrac{1}{4}y^2\right)dy$$

Thus

$$\bar{y} = \frac{\displaystyle\int_0^2 y\left(\tfrac{1}{2}y - \tfrac{1}{4}y^2\right)dy}{1/3}$$

$$= 3\int_0^2 \left(\tfrac{1}{2}y^2 - \tfrac{1}{4}y^3\right)dy$$

$$= 3\left(\tfrac{1}{6}y^3 - \tfrac{1}{16}y^4\right)\Big|_0^2$$

$$= 3\left(\tfrac{4}{3} - 1\right) = 3\left(\tfrac{1}{3}\right) = 1$$

29.

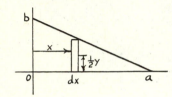

Slope of line: $-\dfrac{b}{a}$; y-intercept: $(0,b)$.

By the slope-intercept formula, $y = -\dfrac{b}{a}x + b$

$$\bar{x} = \frac{M_y}{A} = \frac{\displaystyle\int_0^a x\left(-\tfrac{b}{a}x + b\right)dx}{\tfrac{1}{2}ab} = \frac{2}{ab}\int_0^a \left(-\tfrac{b}{a}x^2 + bx\right)dx$$

$$= \frac{2}{ab}\left(-\tfrac{b}{a}\,\tfrac{x^3}{3} + b\tfrac{x^2}{2}\right)\Big|_0^a = \frac{2}{ab}\left(-\tfrac{b}{a}\,\tfrac{a^3}{3} + b\tfrac{a^2}{2}\right)$$

$$= \frac{2}{ab}\left(-\tfrac{a^2 b}{3} + \tfrac{a^2 b}{2}\right) = \frac{2}{ab}(a^2 b)\left(\tfrac{1}{2} - \tfrac{1}{3}\right) = 2a\left(\tfrac{1}{6}\right) = \tfrac{a}{3}$$

$\bar{y} =$ moment (with respect to x-axis) of typical element:

$$\left(\tfrac{1}{2}y\right)y\,dx = \tfrac{1}{2}y^2 dx$$

$$\bar{y} = \frac{M_x}{A} = \frac{\tfrac{1}{2}\displaystyle\int_0^a \left(-\tfrac{b}{a}x + b\right)^2 dx}{\tfrac{1}{2}ab} = \frac{1}{ab}\int_0^a \left(\tfrac{b^2}{a^2}x^2 - \tfrac{2b^2}{a}x + b^2\right)dx$$

$$= \frac{1}{ab}\left(\tfrac{b^2}{a^2}\,\tfrac{x^3}{3} - \tfrac{2b^2}{a}\,\tfrac{x^2}{2} + b^2 x\right)\Big|_0^a = \frac{1}{ab}\left(\tfrac{b^2}{a^2}\,\tfrac{a^3}{3} - \tfrac{b^2 a^2}{a} + ab^2\right)$$

$$= \frac{1}{ab}\left(\tfrac{1}{3}ab^2 - ab^2 + ab^2\right) = \frac{1}{ab}\left(\tfrac{1}{3}ab^2\right) = \tfrac{b}{3}$$

33.

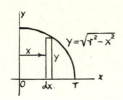

Volume of typical disk: $\pi y^2 dx$

Moment of typical disk: $x(\pi y^2 dx)$

$$= \pi x\left(\sqrt{r^2 - x^2}\right)^2 dx$$

$$\bar{x} = \frac{M_y}{V} = \frac{\pi \int_0^r x\left(\sqrt{r^2 - x^2}\right)^2 dx}{\pi \int_0^r \left(\sqrt{r^2 - x^2}\right)^2 dx} = \frac{\int_0^r x(r^2 - x^2)\,dx}{\int_0^r (r^2 - x^2)\,dx}$$

$$= \frac{\int_0^r (r^2 x - x^3)\,dx}{\int_0^r (r^2 - x^2)\,dx} = \frac{r^2\frac{x^2}{2} - \frac{x^4}{4}\Big|_0^r}{r^2 x - \frac{x^3}{3}\Big|_0^r}$$

$$= \frac{\frac{r^4}{2} - \frac{r^4}{4}}{r^3 - \frac{r^3}{3}} = \frac{r^4\left(\frac{1}{2} - \frac{1}{4}\right)}{r^3\left(1 - \frac{1}{3}\right)} = \frac{r\left(\frac{1}{4}\right)}{\frac{2}{3}} = \frac{3}{8}r$$

$\bar{y} = 0$

37.

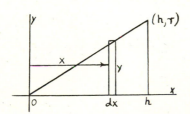

Slope of line: $\frac{r}{h}$

Equation: $y = \frac{r}{h}x$

Moment of typical disk: $x(\pi y^2 dx) = \pi x\left(\frac{r}{h}x\right)^2 dx$

$$\bar{x} = \frac{\pi \int_0^h x\left(\frac{r}{h}x\right)^2 dx}{\pi \int_0^h \left(\frac{r}{h}x\right)^2 dx} = \frac{(r^2/h^2) \int_0^h x^3 dx}{(r^2/h^2) \int_0^h x^2 dx}$$

$$= \frac{\int_0^h x^3 dx}{\int_0^h x^2 dx} = \frac{\frac{x^4}{4}\Big|_0^h}{\frac{x^3}{3}\Big|_0^h} = \frac{h^4}{4}\frac{3}{h^3} = \frac{3}{4}h$$

Since $\bar{y} = 0$, we conclude that the centroid lies on the axis, one-fourth of the way from the base.

1.

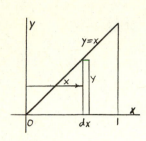

Moment of inertia of typical element: $x^2 \cdot \rho y \, dx = \rho x^2 \cdot x \, dx$

$$I_y = \int_0^1 \rho x^2 x \, dx = \rho \left. \frac{x^4}{4} \right|_0^1 = \frac{\rho}{4}$$

Mass: $\rho \int_0^1 x \, dx = \frac{\rho}{2}$

$$R_y = \sqrt{\frac{\rho}{4} \frac{2}{\rho}} = \sqrt{\frac{1}{2}} = \frac{\sqrt{2}}{2}$$

5.

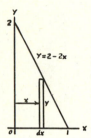

I_y: moment of inertia of typical element: $x^2 \cdot \rho y \, dx$

$$= \rho x^2 \cdot (2 - 2x) \, dx$$

$$I_y = \int_0^1 \rho x^2 (2 - 2x) \, dx = \rho \int_0^1 (2x^2 - 2x^3) \, dx$$

$$= \rho \left(\frac{2}{3} x^3 - \frac{1}{2} x^4 \right) \Big|_0^1 = \rho \left(\frac{2}{3} - \frac{1}{2} \right) = \frac{\rho}{6}$$

Since $A = \frac{1}{2} bh = \frac{1}{2}(1)(2) = 1$, the mass $= \rho$. Thus

$$R_y = \sqrt{\frac{\rho}{6} \frac{1}{\rho}} = \sqrt{\frac{1}{6}} = \frac{\sqrt{6}}{6}$$

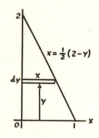

I_x: moment of inertia of typical element:

$$y^2 \cdot \rho x \, dy = \rho y^2 \left[\frac{1}{2}(2 - y) \right] dy$$

$$I_x = \rho \int_0^2 y^2 \left[\frac{1}{2}(2 - y) \right] dy = \frac{\rho}{2} \int_0^2 (2y^2 - y^3) \, dy = \frac{2\rho}{3}$$

$$R_x = \sqrt{\frac{2\rho}{3} \frac{1}{\rho}} = \sqrt{\frac{2}{3}} = \frac{\sqrt{6}}{3}$$

9. Substituting $y = \frac{1}{2}x$ in $y^2 = x$, we get

$$\frac{1}{4}x^2 = x$$
$$x^2 - 4x = 0$$
$$x(x - 4) = 0$$
$$x = 0,\ 4$$

The points of intersection are $(0,0)$ and $(4,2)$.

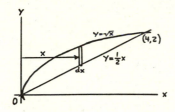

Moment of inertia (with respect to y-axis) of typical element:

$$x^2 \cdot \rho \cdot (\text{height}) \cdot dx = \rho x^2\left(\sqrt{x} - \tfrac{1}{2}x\right)dx$$

$$I_y = \rho \int_0^4 x^2\left(\sqrt{x} - \tfrac{1}{2}x\right)dx = \rho \int_0^4 \left(x^{5/2} - \tfrac{1}{2}x^3\right)dx$$

$$= \rho\left(\tfrac{2}{7}x^{7/2} - \tfrac{1}{8}x^4\right)\Big|_0^4 = \rho\left[\tfrac{2}{7}(4^{7/2}) - \tfrac{1}{8}(4^4)\right] = \rho\left(\tfrac{2}{7} \cdot 128 - 32\right)$$

$$= 32\rho\left(\tfrac{8}{7} - 1\right) = 32\rho\left(\tfrac{8 - 7}{7}\right) = \frac{32\rho}{7}$$

Mass: $\rho \int_0^4 \left(\sqrt{x} - \tfrac{1}{2}x\right)dx = \rho\left(\tfrac{2}{3}x^{3/2} - \tfrac{1}{4}x^2\right)\Big|_0^4 = \rho\left(\tfrac{2}{3} \cdot 8 - 4\right)$

$$= \frac{4\rho}{3}$$

$$R_y = \sqrt{\frac{32\rho}{7}\frac{3}{4\rho}} = \sqrt{\frac{8 \cdot 3}{7}} = \frac{2\sqrt{2} \cdot 3}{\sqrt{7}} = \frac{2\sqrt{42}}{7}$$

13. Solving the equations simultaneously, we get

$$x = y^2 + 2$$
$$\underline{x = y + 2}$$
$$0 = y^2 - y \quad (\text{subtracting})$$
$$y(y - 1) = 0$$
$$y = 0,1$$

The points of intersection are $(2,0)$ and $(3,1)$. Note that the typical element has to be drawn horizontally.

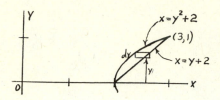

Moment of inertia (with respect to x-axis) of typical element:

$\rho y^2 \cdot (\text{height}) \cdot dy = \rho y^2 [(y + 2) - (y^2 + 2)] dy$

$$I_x = \rho \int_0^1 y^2 [(y + 2) - (y^2 + 2)] dy = \rho \int_0^1 y^2 (y - y^2) dy$$

$$= \rho \int_0^1 (y^3 - y^4) dy = \rho \left(\tfrac{1}{4} y^4 - \tfrac{1}{5} y^5\right)\Big|_0^1 = \rho\left(\tfrac{1}{4} - \tfrac{1}{5}\right) = \tfrac{\rho}{20}$$

Mass: $\rho \int_0^1 [(y + 2) - (y^2 + 2)] dy = \rho \int_0^1 (y - y^2) dy = \tfrac{\rho}{6}$

$$R_x = \sqrt{\tfrac{\rho}{20} \tfrac{6}{\rho}} = \sqrt{\tfrac{3}{10}} = \tfrac{\sqrt{30}}{10}$$

17.

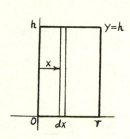

Mass of shell: $\rho (2\pi x y \, dx)$

Moment of inertia of typical shell (since $y = h$):

$x^2 \cdot \rho (2\pi x y \, dx) = 2\pi \rho x^3 h \, dx$

$$I_y = 2\pi \rho \int_0^r x^3 h \, dx = 2\pi \rho h \left.\tfrac{x^4}{4}\right|_0^r$$

$$= \tfrac{1}{2} \pi r^4 h \rho$$

Since the mass of the cylinder is $\rho \pi r^2 h$, I_y can also be written as follows:

$$I_y = \tfrac{1}{2} (\rho \pi r^2 h) r^2 = \tfrac{1}{2} m r^2$$

$$R_y = \sqrt{\tfrac{\pi r^4 h \rho}{2} \tfrac{1}{\rho \pi r^2 h}} = \sqrt{\tfrac{r^2}{2}} = \tfrac{r}{\sqrt{2}} = \tfrac{r\sqrt{2}}{2}$$

21.

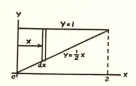

Mass of shell: $\rho \cdot 2\pi$ (radius) $\cdot$ (height) $\cdot$ (thickness)

$$= \rho \cdot 2\pi \cdot x \cdot \left(1 - \tfrac{1}{2}x\right)dx$$

$$= 2\pi\rho x\left(1 - \tfrac{1}{2}x\right)dx$$

Moment of inertia of typical shell:

$$x^2 \cdot 2\pi\rho x\left(1 - \tfrac{1}{2}x\right)dx$$

$$I_y = \int_0^2 x^2 \cdot 2\pi\rho x\left(1 - \tfrac{1}{2}x\right)dx = 2\pi\rho \int_0^2 \left(x^3 - \tfrac{1}{2}x^4\right)dx$$

$$= 2\pi\rho\left(\frac{x^4}{4} - \frac{x^5}{10}\right)\Big|_0^2 = 2\pi\rho\left(4 - \frac{16}{5}\right) = 2\pi\rho\left(\frac{4}{5}\right)$$

$$= \frac{8\pi\rho}{5}$$

25.

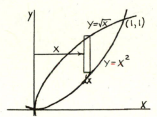

Mass of shell:

$$\rho \cdot 2\pi x \cdot \text{(height)} \cdot dx$$
$$= \rho \cdot 2\pi x(\sqrt{x} - x^2)\,dx$$

Moment of inertia of shell:

$$x^2 \cdot \rho \cdot 2\pi x(\sqrt{x} - x^2)\,dx$$
$$= 2\pi\rho x^3(\sqrt{x} - x^2)\,dx$$

$$I_y = 2\pi\rho \int_0^1 x^3(\sqrt{x} - x^2)\,dx = 2\pi\rho \int_0^1 (x^{7/2} - x^5)\,dx$$

$$= 2\pi\rho\left(\frac{2}{9}x^{9/2} - \frac{1}{6}x^6\right)\Big|_0^1 = 2\pi\rho\left(\frac{2}{9} - \frac{1}{6}\right) = 2\pi\rho\left(\frac{4}{18} - \frac{3}{18}\right) = \frac{\pi\rho}{9}$$

Mass: $2\pi\rho \int_0^1 x(\sqrt{x} - x^2)\,dx = 2\pi\rho \int_0^1 (x^{3/2} - x^3)\,dx$

$$= 2\pi\rho\left(\frac{2}{5}x^{5/2} - \frac{1}{4}x^4\right)\Big|_0^1$$

$$= 2\pi\rho\left(\frac{2}{5} - \frac{1}{4}\right) = \frac{3\pi\rho}{10}$$

$$R_y = \sqrt{\frac{\pi\rho}{9} \cdot \frac{10}{3\pi\rho}} = \sqrt{\frac{10}{9 \cdot 3}} = \frac{\sqrt{10}}{3\sqrt{3}} = \frac{\sqrt{30}}{9}$$

Section 5.6

1. From Hooke's law,

$$F = kx$$

$$6 = k \cdot \frac{1}{8} \qquad \text{or} \qquad k = 48$$

So F = 48x. Since the spring is stretched 2 ft,

$$W = \int_0^2 48x\,dx = 24x^2\big|_0^2 = 24 \cdot 4 = 96 \text{ ft-lb}$$

5.

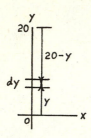

Weight of typical element:

$$3 \text{ N/m} \times dy \text{ m} = 3dy \text{ N}$$

Work done in moving the typical element to the top:

$$3dy \cdot (20 - y) = 3(20 - y)\,dy$$

Summing from y = 0 to y = 20, we get

$$W = \int_0^{20} 3(20 - y)\,dy = 3\left(20y - \tfrac{1}{2}y^2\right)\Big|_0^{20} = 600 \text{ J}$$

9.

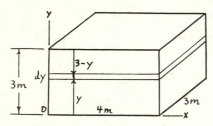

Volume of typical element: $4 \cdot 3 \cdot dy = 12dy$

Weight of typical element: $(12dy)w = 12wdy$

Work done in moving the typical element to the top: $(3 - y) \cdot 12wdy$

Summing from y = 0 to y = 3, we get

$$W = \int_0^3 (3 - y) \cdot 12wdy = 12w\left(3y - \tfrac{1}{2}y^2\right)\Big|_0^3$$

$$= 12w\left(9 - \tfrac{9}{2}\right) = 54w \text{ J}$$

108

13.

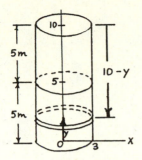

Volume of typical element: $\pi(3^2)\,dy = 9\pi\,dy$

Weight of typical element: $9\pi\,dy \cdot w = 9\pi w\,dy$

Work done in moving the typical element to the top: $(10 - y) \cdot 9\pi w\,dy$

Summing from $y = 0$ to $y = 5$, we get

$$W = \int_0^5 (10 - y) \cdot 9\pi w\,dy = 9\pi w \int_0^5 (10 - y)\,dy$$

$$= 9\pi w \left(10y - \tfrac{1}{2}y^2\right)\Big|_0^5 = 9\pi w \left(50 - \tfrac{25}{2}\right)$$

$$= \tfrac{675\pi w}{2} \ J$$

17.

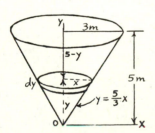

To find the volume and weight of the typical element, we need to determine the radius x. Since the line on the right has slope $m = \tfrac{5}{3}$, its equation is $y = \tfrac{5}{3}x$, or $x = \tfrac{3}{5}y$. Thus:

Weight of typical element: $\pi\left(\tfrac{3}{5}y\right)^2\,dy \cdot w$

Work done in moving the typical element to the top:

$$(5 - y) \cdot \pi\left(\tfrac{3}{5}y\right)^2\,dy \cdot w = \tfrac{9\pi w}{25}(5y^2 - y^3)\,dy$$

Summing from $y = 0$ to $y = 5$, we get

$$W = \int_0^5 \frac{9\pi w}{25}(5y^2 - y^3) = \frac{9\pi w}{25}\left(\frac{5y^3}{3} - \frac{y^4}{4}\right)\Big|_0^5$$

$$= \frac{9\pi w}{25}\left(\frac{5^4}{3} - \frac{5^4}{4}\right) = \frac{9\pi w}{25} \cdot 5^4\left(\frac{1}{3} - \frac{1}{4}\right)$$

$$= 9\pi w(25)\frac{1}{12} = \frac{75\pi w}{4} \text{ J}$$

21.

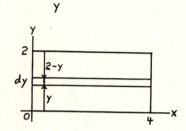

Pressure on strip: $(2 - y)w$

Area of strip: $4\,dy$

Force against strip: $(2 - y)w \cdot 4\,dy = 4w(2 - y)\,dy$

Summing from $y = 0$ to $y = 2$:

$$F = \int_0^2 4w(2 - y)\,dy = 4w\left(2y - \frac{1}{2}y^2\right)\Big|_0^2 = 8w \text{ N}$$

25.

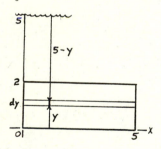

Pressure on strip: $(5 - y)w$

Area of strip: $5\,dy$

Force against strip:

$$(5 - y)w \cdot 5\,dy$$
$$= 5w(5 - y)\,dy$$

Summing from $y = 0$ to $y = 2$:

$$F = \int_0^2 5w(5 - y)\,dy = 5w\left(5y - \frac{1}{2}y^2\right)\Big|_0^2 = 40w \text{ N}$$

29.

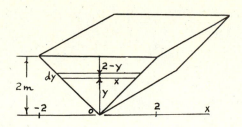

Since the slope of the line on the right is 1, the equation is $y = x$.

Area of strip: $2x \, dy = 2y \, dy$

Pressure on strip: $(2 - y)w$

Force against strip: $(2 - y)w \cdot 2y \, dy = 2w(2y - y^2) \, dy$

Summing from $y = 0$ to $y = 2$, we get

$$F = \int_0^2 2w(2y - y^2) \, dy = 2w\left(y^2 - \tfrac{1}{3}y^3\right)\Big|_0^2$$

$$= 2w\left(4 - \tfrac{8}{3}\right) = \tfrac{8w}{3} \text{ N}$$

33.

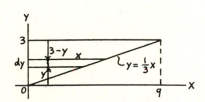

Equation of line: $y = \tfrac{1}{3}x$, or $x = 3y$

Area of strip: $x \, dy = 3y \, dy$

Pressure on strip: $(3 - y)w$

Force against strip: $(3 - y)w \cdot 3y \, dy$

Summing from $y = 0$ to $y = 3$:

$$F = \int_0^3 (3 - y)w \cdot 3y \, dy = 3w \int_0^3 (3y - y^2) \, dy$$

$$= 3w\left(3\tfrac{y^2}{2} - \tfrac{y^3}{3}\right)\Big|_0^3 = 3w\left(\tfrac{3^3}{2} - \tfrac{3^3}{3}\right)$$

$$= 3^4 w\left(\tfrac{1}{2} - \tfrac{1}{3}\right) = 81w\left(\tfrac{1}{6}\right) = \tfrac{27w}{2} \text{ N}$$

37.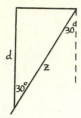

If the end of the trough is tilted, then a point which is z units from the top of the trough is at depth d, as shown in the diagram.

It follows that

$$d = z \cos 30° = \frac{\sqrt{3}}{2} z$$

The other calculations follow Exercise 29:

Pressure on strip: $\frac{\sqrt{3}}{2}(2 - y)w$

Force against strip: $\frac{\sqrt{3}}{2}(2 - y)w \cdot 2y\,dy$

$$F = \int_0^2 \frac{\sqrt{3}}{2}(2 - y)w \cdot 2y\,dy = \frac{\sqrt{3}}{2} \int_0^2 2w(2y - y^2)\,dy$$

$$= \frac{\sqrt{3}}{2}\left(\frac{8w}{3}\right) \text{ N by Exercise 29.}$$

REVIEW EXERCISES FOR CHAPTER 5

1. Taking the upward direction to be positive, $g = -32$. So

$$v = -32t \qquad\qquad v = 0 \text{ when } t = 0$$

and

$$s = -16t^2 + k$$

Since $s = 256$ when $t = 0$, we have

$$256 = 0 + k$$

so that

$$s = -16t^2 + 256$$

So see how long it takes for the object to reach the ground, we let $s = 0$ and solve for t:

$$0 = -16t^2 + 256 \quad \text{ or } \quad t = 4 \text{ s}$$

Taking v to be $32t$, we get

$$v_{av} = \frac{1}{4 - 0} \int_0^4 32t\,dt = \tfrac{1}{4}(16t^2)\big|_0^4 = 64 \text{ ft/s}$$

5. See Exercise 25, Section 5.4.

9.

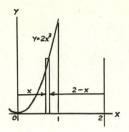

radius of shell = 2 - x

Volume of typical shell: 2π (radius) $\cdot$ (height) $\cdot$ (thickness)
$$= 2\pi (2 - x)(2x^2)\,dx$$

$$V = \int_0^1 2\pi (2 - x)(2x^2)\,dx = 4\pi \int_0^1 (2 - x)(x^2)\,dx$$

$$= 4\pi \int_0^1 (2x^2 - x^3)\,dx = 4\pi \left(\tfrac{2}{3}x^3 - \tfrac{1}{4}x^4\right)\Big|_0^1 = 4\pi \left(\tfrac{2}{3} - \tfrac{1}{4}\right)$$

$$= 4\pi \,\frac{8 - 3}{12} = \frac{5\pi}{3}$$

13.

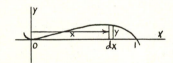

From $y = x^2 - x^3 = x^2(1 - x) = 0$, the x-intercepts are 0 and 1.

$\bar{x}$:   moment (with respect to y-axis) of typical element: $xy\,dx$

$$\bar{x} = \frac{M_y}{A} = \frac{\displaystyle\int_0^1 x(x^2 - x^3)\,dx}{\displaystyle\int_0^1 (x^2 - x^3)\,dx} = \frac{\tfrac{1}{4}x^4 - \tfrac{1}{5}x^5\big|_0^1}{\tfrac{1}{3}x^3 - \tfrac{1}{4}x^4\big|_0^1}$$

$$= \frac{3}{5}$$

$\bar{y}$:   moment (with respect to x-axis) of typical element: $\left(\tfrac{1}{2}y\right)(y\,dx)$

$$\bar{y} = \frac{M_x}{A} = \frac{\displaystyle\int_0^1 \left(\tfrac{1}{2}y\right)(y\,dx)}{A} = \frac{\tfrac{1}{2}\displaystyle\int_0^1 (x^2 - x^3)(x^2 - x^3)\,dx}{1/12}$$

$$= 12\left(\tfrac{1}{2}\right)\int_0^1 (x^4 - 2x^5 + x^6)\,dx$$

$$= 6\left(\tfrac{1}{5}x^5 - \tfrac{1}{3}x^6 + \tfrac{1}{7}x^7\right)\Big|_0^1 = 6\left(\tfrac{1}{5} - \tfrac{1}{3} + \tfrac{1}{7}\right)$$

$$= 6\left(\frac{21 - 35 + 15}{105}\right) = \frac{6}{105} = \frac{2}{35}$$

113

17.

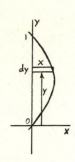

Moment of inertia (with respect to x-axis) of typical element:

$$y^2 \cdot \rho x\,dy = \rho y^2(y - y^2)\,dy$$

$$I_x = \rho \int_0^1 y^2(y - y^2)\,dy = \rho\left(\tfrac{1}{4}y^4 - \tfrac{1}{5}y^5\right)\Big|_0^1$$

$$= \rho\left(\tfrac{1}{4} - \tfrac{1}{5}\right) = \tfrac{\rho}{20}$$

Mass: $\rho \displaystyle\int_0^1 (y - y^2)\,dy = \tfrac{1}{2}y^2 - \tfrac{1}{3}y^3\Big|_0^1 \rho = \tfrac{\rho}{6}$

$$R_y = \sqrt{\tfrac{\rho}{20}\tfrac{6}{\rho}} = \sqrt{\tfrac{3}{10}} = \tfrac{\sqrt{30}}{10}$$

21. By Hooke's law, $F = kx$, $2 = k \cdot \tfrac{1}{2}$, or $k = 4$. Thus $F = 4x$.

(a) The spring is stretched 2 ft beyond its natural length. Thus

$$W = \int_0^2 4x\,dx = 2x^2\Big|_0^2 = 8 \text{ ft-lb}$$

(b) $W = \displaystyle\int_1^3 4x\,dx = 2x^2\Big|_1^3 = 16 \text{ ft-lb}$

25.

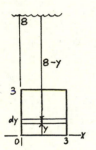

Pressure on strip: $(8 - y)w$

Area of strip: $3\,dy$

Force against strip: $(8 - y)w \cdot 3\,dy$

Summing from $y = 0$ to $y = 3$, we get

$$F = \int_0^3 (8 - y)w \cdot 3\,dy = 3w \int_0^3 (8 - y)\,dy$$

$$= 3w\left(8y - \tfrac{1}{2}y^2\right)\Big|_0^3 = 3w\left(24 - \tfrac{9}{2}\right) = \tfrac{117w}{2} \text{ N}$$

$$= \tfrac{117}{2}(10,000) = 585,000 \text{ N}$$

CHAPTER 6   TRANSCENDENTAL FUNCTIONS

Section 6.1

3.

$$\tan(-45°) = -1$$

5.

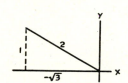

$$\sec 150° = \frac{2}{-\sqrt{3}} = -\frac{2\sqrt{3}}{3}$$

9.

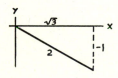

$$\csc(-30°) = -2$$

13.

$$\tan 90° = \frac{1}{0} \quad (\text{undefined})$$

25. $60° = 60° \cdot \dfrac{\pi}{180°} = \dfrac{\pi}{3}$

29. $135° = 135° \cdot \dfrac{\pi}{180°} = \dfrac{3\pi}{4}$

33. $20° = 20° \cdot \dfrac{\pi}{180°} = \dfrac{\pi}{9}$

37. $\dfrac{\pi}{6} = \dfrac{\pi}{6} \cdot \dfrac{180°}{\pi} = 30°$

41. $\dfrac{11\pi}{10} = \dfrac{11\pi}{10} \cdot \dfrac{180°}{\pi} = 198°$

45. amplitude $= \dfrac{1}{3}$ (coefficient of sin 2x)
   period $= \dfrac{2\pi}{2} = \pi$
   (see drawing in answer section)

49. $\cos\theta \, \tan\theta = \cos\theta \, \dfrac{\sin\theta}{\cos\theta} = \sin\theta$

53. Since $1 + \tan^2\theta = \sec^2\theta$, we have $\tan^2\theta - \sec^2\theta = -1$.   So
   $$\dfrac{\tan^2\theta - \sec^2\theta}{\sec\theta} = \dfrac{-1}{\sec\theta} = -\cos\theta$$

57. $\dfrac{1}{\sec^2\theta + \tan^2\theta + \cos^2\theta}$

   $= \dfrac{1}{(\sin^2\theta + \cos^2\theta) + \tan^2\theta}$ rearranging

   $= \dfrac{1}{1 + \tan^2\theta}$ $\sin^2\theta + \cos^2\theta = 1$

   $= \dfrac{1}{\sec^2\theta}$ $1 + \tan^2\theta = \sec^2\theta$

   $= \cos^2\theta$ $\cos\theta = \dfrac{1}{\sec\theta}$

61. $\cos\theta \, \cot\theta + \sin\theta$

   $= \cos\theta \, \dfrac{\cos\theta}{\sin\theta} + \sin\theta$ $\cot\theta = \dfrac{\cos\theta}{\sin\theta}$

   $= \dfrac{\cos^2\theta}{\sin\theta} + \dfrac{\sin^2\theta}{\sin\theta}$ common denominator $= \sin\theta$

   $= \dfrac{\cos^2\theta + \sin^2\theta}{\sin\theta}$

   $= \dfrac{1}{\sin\theta}$ $\cos^2\theta + \sin^2\theta = 1$

81. $\cos\left(x - \dfrac{\pi}{2}\right) = \cos x \, \cos\dfrac{\pi}{2} + \sin x \, \sin\dfrac{\pi}{2}$

   $= \cos x \cdot 0 + \sin x \cdot 1 = \sin x$

85. Since $\sin 2\theta = 2\sin\theta \cos\theta$,

$$2 \sin 5x \cos 5x = \sin(2 \cdot 5x) = \sin 10x$$

89. Since $\cos 2\theta = \cos^2\theta - \sin^2\theta$,

$$\cos^2 3x - \sin^2 3x = \cos(2 \cdot 3x) = \cos 6x$$

93. Since $\sin^2\theta = \frac{1}{2}(1 - \cos 2\theta)$,

$$\sin^2 3x = \frac{1}{2}[1 - \cos(2 \cdot 3x)] = \frac{1}{2}(1 - \cos 6x)$$

97. $\sin^2 \frac{1}{2}x = \frac{1}{2}[1 - \cos(2 \cdot \frac{1}{2}x)] = \frac{1}{2}(1 - \cos x)$

Section 6.2

1. $y = \cos 5x$

$$y' = (-\sin 5x)\frac{d}{dx}(5x) = (-\sin 5x)(5) = -5 \sin 5x$$

5. $y = \sin x^2$

$$y' = (\cos x^2)\frac{d}{dx}(x^2) = (\cos x^2)(2x) = 2x \cos x^2$$

11. $y = x \sin x$. By the product rule,

$$y' = x\,\frac{d}{dx} \sin x + (\sin x)\frac{d}{dx}(x) = x \cos x + \sin x$$

13. $y = \frac{\sin x}{x}$. By the quotient rule,

$$y' = \frac{x(d/dx)\sin x - \sin x(dx/dx)}{x^2} = \frac{x \cos x - \sin x}{x^2}$$

17. $y = x \cos 2x$. By the product rule,

$$y' = x\frac{d}{dx} \cos 2x + \cos 2x \,\frac{d}{dx}(x)$$

$$= x(-\sin 2x)\frac{d}{dx}(2x) + \cos 2x \cdot 1$$

$$= x(-\sin 2x)(2) + \cos 2x$$

$$= -2x \sin 2x + \cos 2x$$

$$= \cos 2x - 2x \sin 2x$$

21. $y = \sin \frac{1}{x}$

$$y' = \left(\cos \tfrac{1}{x}\right)\frac{d}{dx}\left(\tfrac{1}{x}\right) = \left(\cos \tfrac{1}{x}\right)\frac{d}{dx}(x^{-1}) = \left(\cos \tfrac{1}{x}\right)(-1x^{-2})$$

$$= -\frac{\cos(1/x)}{x^2}$$

117

25. $y = \frac{x}{\sin 4x}$. By the quotient rule,

$$y' = \frac{(\sin 4x)(d/dx)x - x(d/dx)\sin 4x}{(\sin 4x)^2}$$

$$= \frac{(\sin 4x) \cdot 1 - x \cos 4x \cdot 4}{\sin^2 4x}$$

$$= \frac{\sin 4x - 4x \cos 4x}{\sin^2 4x}$$

29. $y = \sqrt{x} \sin x$. By the product rule,

$$y' = x^{1/2}\frac{d}{dx}\sin x + \sin x\frac{d}{dx}(x^{1/2}) = x^{1/2}\cos x + (\sin x)\tfrac{1}{2}x^{-1/2}$$

$$= \sqrt{x} \cos x + \frac{1}{2\sqrt{x}} \sin x$$

33. $y = \sin x \cos x$. By the product rule,

$$y' = (\sin x)\frac{d}{dx}\cos x + (\cos x)\frac{d}{dx}\sin x$$

$$= \sin x(-\sin x) + \cos x(\cos x) = -\sin^2 x + \cos^2 x$$
$$= \cos^2 x - \sin^2 x = \cos 2x$$
by the double-angle formula.

37. $y = x \cos^2 3x = x(\cos 3x)^2$. By the product rule,

$$y' = x\,\frac{d}{dx}(\cos 3x)^2 + (\cos 3x)^2\,\frac{d}{dx}(x)$$

$$= x\,\frac{d}{dx}(\cos 3x)^2 + \cos^2 3x \cdot 1$$

Next, by the generalized power rule,

$$y' = x \cdot 2(\cos 3x)\frac{d}{dx}\cos 3x + \cos^2 3x$$

$$= x \cdot 2(\cos 3x)(-\sin 3x \cdot 3) + \cos^2 3x$$
$$= -6x \cos 3x \sin 3x + \cos^2 3x$$

After factoring cos 3x, we get
$$y' = \cos 3x(\cos 3x - 6x \sin 3x)$$

41. $y = \sin x$, $y' = \cos x$, $y'' = -\sin x$, $y^{(3)} = -\cos x$, and $y^{(4)} = \sin x$.
Thus
$$\frac{d^4}{dx^4} \sin x = \sin x$$

Section 6.3

1. $y = \sec 5x$

$$y' = (\sec 5x \tan 5x)\frac{d}{dx}(5x) = (\sec 5x \tan 5x)(5) = 5 \sec 5x \tan 5x$$

118

5. $y = 3 \cot 4x$

$y' = 3 \frac{d}{dx} \cot 4x = 3(-\csc^2 4x) \frac{d}{dx}(4x) = 3(-\csc^2 4x)(4) = -12 \csc^2 4x$

9. $s = \tan 2t$

$\frac{ds}{dt} = (\sec^2 2t) \frac{d}{dt}(2t) = (\sec^2 2t)(2) = 2 \sec^2 2t$

13. $y = \sec(x^3 + 1)$

$y' = [\sec(x^3 + 1) \tan(x^3 + 1)] \frac{d}{dx}(x^3 + 1)$

$= 3x^2 \sec(x^3 + 1) \tan(x^3 + 1)$

17. $y = \cot \sqrt{3x} = \cot(3x)^{1/2}$

$y' = [-\csc^2(3x)^{1/2}] \frac{d}{dx}(3x)^{1/2} = [-\csc^2(3x)^{1/2}] \frac{1}{2}(3x)^{-1/2}(3)$

$= \frac{-3 \csc^2(3x)^{1/2}}{2(3x)^{1/2}} = -\frac{3 \csc^2 \sqrt{3x}}{2\sqrt{3x}}$

19. $y = \sqrt{\tan 2x} = (\tan 2x)^{1/2}$.   By the power rule,

$y' = \frac{1}{2}(\tan 2x)^{-1/2} \frac{d}{dx} \tan 2x = \frac{1}{2}(\tan 2x)^{-1/2}(\sec^2 2x)(2) = \frac{\sec^2 2x}{\sqrt{\tan 2x}}$

21. $y = 2 \tan^4 4x = 2(\tan 4x)^4$.   By the power rule,

$y' = 2 \cdot 4(\tan 4x)^3 \frac{d}{dx} \tan 4x$

$= 8(\tan 4x)^3 \sec^2 4x \cdot 4$

$= 32 \tan^3 4x \sec^2 4x$

25. $T_1 = T_2^2 \csc T_2$.   By the product rule,

$\frac{dT_1}{dT_2} = T_2^2 \frac{d}{dT_2} \csc T_2 + \csc T_2 \frac{d}{dT_2}(T_2^2)$

$= T_2^2(-\csc T_2 \cot T_2) + \csc T_2(2T_2)$

$= -T_2^2 \csc T_2 \cot T_2 + 2T_2 \csc T_2$

$= T_2 \csc T_2(2 - T_2 \cot T_2)$

29. $y = \frac{1 + \tan x}{\sin x}$.   By the quotient rule,

$y' = \frac{(\sin x)(d/dx)(1 + \tan x) - (1 + \tan x)(d/dx)(\sin x)}{(\sin x)^2}$

$= \frac{(\sin x)(\sec^2 x) - (1 + \tan x)(\cos x)}{\sin^2 x}$

$= \frac{\sin x \sec^2 x - \cos x - \tan x \cos x}{\sin^2 x}$

Since $\tan x \cos x = \frac{\sin x}{\cos x} \cos x = \sin x$, we get

$$y' = \frac{\sin x \sec^2 x - \cos x - \sin x}{\sin^2 x}$$

33. $y = \frac{x^3}{\tan 3x}$. By the quotient rule,

$$y' = \frac{\tan 3x (d/dx) x^3 - x^3 (d/dx) \tan 3x}{(\tan 3x)^2}$$

$$= \frac{\tan 3x (3x^2) - x^3 (\sec^2 3x \cdot 3)}{\tan^2 3x}$$

$$= \frac{3x^2 \tan 3x - 3x^3 \sec^2 3x}{\tan^2 3x}$$

$$= \frac{3x^2 (\tan 3x - x \sec^2 3x)}{\tan^2 3x}$$

37. By the power rule,

$$\frac{d}{dx} \tan^3 x = (3 \tan^2 x) \frac{d}{dx} \tan x = 3 \tan^2 x \sec^2 x$$

Thus

$$\frac{d}{dx}\left(\frac{1}{3} \tan^3 x + \tan x\right) = \frac{1}{3} \cdot 3 \tan^2 x \sec^2 x + \sec^2 x$$

$$= \sec^2 x (\tan^2 x + 1) \qquad \text{factoring}$$

$$= \sec^2 x \sec^2 x \qquad\qquad 1 + \tan^2 x = \sec^2 x$$

$$= \sec^4 x$$

41. $y = x \tan x$. By the product rule,
$y' = x \sec^2 x + \tan x$. By the product and power rules,

$$y'' = x \frac{d}{dx}(\sec x)^2 + (\sec^2 x) \cdot 1 + \sec^2 x$$

$$= x(2 \sec x)(\sec x \tan x) + \sec^2 x + \sec^2 x$$

$$= 2x \sec^2 x \tan x + 2 \sec^2 x$$

$$= 2 \sec^2 x (x \tan x + 1)$$

45. Treating y as a function of x, we get

$$\frac{d}{dx}(y) = \frac{dy}{dx} \quad \text{and} \quad \frac{d}{dx}(y^2) = 2y \frac{dy}{dx}:$$

$$y^2 = x \sec x$$

$$2y \frac{dy}{dx} = x \sec x \tan x + \sec x$$

$$\frac{dy}{dx} = \frac{x \sec x \tan x + \sec x}{2y}$$

$$= \frac{\sec x (x \tan x + 1)}{2y}$$

120

49. $y = x \cot y^2$

$\dfrac{dy}{dx} = x \dfrac{d}{dx} \cot y^2 + \cot y^2 \cdot 1$ product rule

$\dfrac{dy}{dx} = x(-\csc^2 y^2)\left(2y \dfrac{dy}{dx}\right) + \cot y^2$ $\dfrac{d}{dx} y^2 = 2y \dfrac{dy}{dx}$

$\dfrac{dy}{dx} + (2xy \csc^2 y^2) \dfrac{dy}{dx} = \cot y^2$

$\dfrac{dy}{dx}(1 + 2xy \csc^2 y^2) = \cot y^2$ factoring $\dfrac{dy}{dx}$

$\dfrac{dy}{dx} = \dfrac{\cot y^2}{1 + 2xy \csc^2 y^2}$

53. Slope of tangent line:

$\dfrac{dy}{dx} = 2(-\csc^2 2x) \cdot 2 = -4 \csc^2 2x \big|_{x=\pi/8}$

$\qquad = -4\left(\csc \tfrac{\pi}{4}\right)^2 = -4(\sqrt{2})^2 = -8$

Slope of normal line: $-\dfrac{1}{-8} = \dfrac{1}{8}$

Section 6.4

17. Let $\theta = \text{Arctan } 6 = \text{Arctan } \dfrac{6}{1}$. We place 6 on the side opposite θ and 1 on the side adjacent. Since $\sqrt{6^2 + 1^2} = \sqrt{37}$, the length of the hypotenuse, we get

$\cos(\text{Arctan } 6) = \cos \theta = \dfrac{1}{\sqrt{37}} = \dfrac{\sqrt{37}}{37}$

21.

$\sec\left[\text{Arcsin}\left(-\tfrac{1}{3}\right)\right] = \sec \theta = \dfrac{3}{2\sqrt{2}} = \dfrac{3\sqrt{2}}{4}$

25.

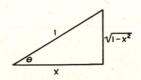

Let θ = Arccos $\frac{x}{1}$. By the Pythagorean theorem, the length of the remaining side is $\sqrt{1 - x^2}$. Thus $\sin(\text{Arccos } x) = \sin \theta = \sqrt{1 - x^2}$

29.

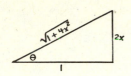

Let θ = Arctan $\frac{2x}{1}$.

Length of hypotenuse:

$$\sqrt{1^2 + (2x)^2} = \sqrt{1 + 4x^2}$$

$$\cos(\text{Arctan } 2x) = \cos \theta = \frac{1}{\sqrt{1 + 4x^2}}$$

33.

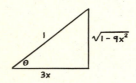

Let θ = Arccos $\frac{3x}{1}$. By the Pythagorean theorem, the length of the remaining side is $\sqrt{1 - 9x^2}$. Thus

$$\sin(\text{Arccos } 3x) = \sin \theta = \frac{\sqrt{1 - 9x^2}}{1}$$

37.

$$y = 1 + \sin x$$
$$\sin x = y - 1$$
$$x = \text{Arcsin}(y - 1)$$

41.

$$y = 3 \tan 4x + 1$$
$$y - 1 = 3 \tan 4x$$
$$\frac{y - 1}{3} = \tan 4x$$
$$\tan 4x = \frac{y - 1}{3}$$
$$4x = \text{Arctan } \frac{y - 1}{3}$$
$$x = \frac{1}{4} \text{Arctan } \frac{y - 1}{3}$$

Section 6.5

1. $y = \text{Arctan } 3x$

$$y' = \frac{1}{1 + (3x)^2} \frac{d}{dx}(3x) = \frac{3}{1 + 9x^2}$$

5. $y = \text{Arctan } 2x^2$

$$y' = \frac{1}{1 + (2x^2)^2} \frac{d}{dx}(2x^2) = \frac{4x}{1 + 4x^4}$$

9. $y = \text{Arcsin } 2w$

$$\frac{dy}{dw} = \frac{1}{\sqrt{1 - (2w)^2}} \frac{d}{dw}(2w) = \frac{2}{\sqrt{1 - 4w^2}}$$

122

13. $y = \text{Arcsin } 2x^2$

$$y' = \frac{1}{\sqrt{1 - (2x^2)^2}} \frac{d}{dx}(2x^2) = \frac{4x}{\sqrt{1 - 4x^4}}$$

17. $y = x \text{ Arctan } x$. By the product rule,

$$y' = x \frac{d}{dx} \text{Arctan } x + \text{Arctan } x \cdot 1 = \frac{x}{1 + x^2} + \text{Arctan } x$$

21. $r = \theta \text{ Arcsin } 3\theta$. By the product rule,

$$\frac{dr}{d\theta} = \theta \frac{d}{d\theta} \text{Arcsin } 3\theta + \text{Arcsin } 3\theta \cdot 1$$

$$= \frac{\theta}{\sqrt{1 - (3\theta)^2}} \frac{d}{d\theta}(3\theta) + \text{Arcsin } 3\theta$$

$$= \frac{3\theta}{\sqrt{1 - 9\theta^2}} + \text{Arcsin } 3\theta$$

25. $y = \frac{\text{Arcsin } x}{x}$. By the quotient rule,

$$y' = \frac{x \frac{d}{dx} \text{Arcsin } x - \text{Arcsin } x \cdot 1}{x^2}$$

$$= \frac{\frac{x}{\sqrt{1 - x^2}} - \text{Arcsin } x}{x^2}$$

Now simplify the complex fraction by multiplying numerator and denominator by $\sqrt{1 - x^2}$:

$$= \frac{\frac{x}{\sqrt{1 - x^2}} - \text{Arcsin } x}{x^2} \cdot \frac{\sqrt{1 - x^2}}{\sqrt{1 - x^2}} = \frac{x - \sqrt{1 - x^2} \text{ Arcsin } x}{x^2\sqrt{1 - x^2}}$$

29. $y = x^{1/2} \text{Arcsin } x$. By the product rule,

$$y' = x^{1/2} \frac{d}{dx} \text{Arcsin } x + (\text{Arcsin } x)\frac{d}{dx}(x^{1/2})$$

$$= x^{1/2} \frac{1}{\sqrt{1 - x^2}} + (\text{Arcsin } x)\left(\tfrac{1}{2}x^{-1/2}\right) = \frac{\sqrt{x}}{\sqrt{1 - x^2}} + \frac{\text{Arcsin } x}{2\sqrt{x}}$$

33. $y = (\text{Arccos } x)^{1/2}$. By the power rule,

$$y' = \tfrac{1}{2}(\text{Arccos } x)^{-1/2}\left(-\frac{1}{\sqrt{1 - x^2}}\right)$$

$$= -\tfrac{1}{2}(\text{Arccos } x)^{-1/2}(1 - x^2)^{-1/2} = -\tfrac{1}{2}[(1 - x^2)\text{Arccos } x]^{-1/2}$$

37. Let $y = \text{Arccot } u$, so that $u = \cot y$. Then

$$\frac{du}{dx} = -\csc^2 y \frac{dy}{dx}$$

$$\frac{dy}{dx} = -\frac{1}{\csc^2 y} \frac{du}{dx} = -\frac{1}{1 + \cot^2 y} \frac{du}{dx} = -\frac{1}{1 + u^2} \frac{du}{dx}$$

123

1. $3^3 = 27$; base: 3; exponent: 3. Thus $\log_3 27 = 3$

5. $(32)^{-1/5} = \frac{1}{2}$; base: 32; exponent: $-\frac{1}{5}$. Thus $\log_{32} \frac{1}{2} = -\frac{1}{5}$

9. $\log_{1/4} \frac{1}{16} = 2$; base: $\frac{1}{4}$; exponent: 2. Thus $\left(\frac{1}{4}\right)^2 = \frac{1}{16}$

13. $\log_{1/3} x = 2$. Changing to exponential form, we get

$$\left(\frac{1}{3}\right)^2 = x \quad \text{or} \quad x = \frac{1}{9}$$

17. $\log_x \frac{1}{3} = -\frac{1}{3}$

$x^{-1/3} = \frac{1}{3}$ exponential form

$\frac{1}{x^{1/3}} = \frac{1}{3}$

$x^{1/3} = 3$

$\left(x^{1/3}\right)^3 = 3^3$

$x = 27$

21.

x:	−1	0	1	2
y:	$\frac{1}{3}$	1	3	9

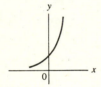

25. $y = \log_3 x \quad \text{or} \quad x = 3^y$

y:	−1	0	1	2
x:	$\frac{1}{3}$	1	3	9

29. $5 \log_5 2 - 3 \log_5 2$

$\quad = \log_5 2^5 - \log_5 2^3$ $\qquad\qquad$ $k \; \log_a M = \log_a M^k$

$\quad = \log_5 32 - \log_5 8$

$\quad = \log_5 \dfrac{32}{8} = \log_5 4$ $\qquad$ $\log_a M - \log_a N = \log_a \dfrac{M}{N}$

33. $2 \log_3 y + \dfrac{1}{3} \log_3 8 - 2 \log_3 5$

$\quad = \log_3 y^2 + \log_3 8^{1/3} - \log_3 5^2$ $\qquad$ $k \; \log_a M = \log_a M^k$

$\quad = \log_3 \dfrac{y^2 \cdot 8^{1/3}}{5^2}$ $\qquad\qquad$ properties (6.16) and (6.17)

$\quad = \log_3 \dfrac{2y^2}{25}$

37. $\log_6 \sqrt{6x} = \log_6 (6x)^{1/2}$

$\quad = \dfrac{1}{2} \log_6 6x$ $\qquad\qquad$ $\log_a M^k = k \; \log_a M$

$\quad = \dfrac{1}{2}(\log_6 6 + \log_6 x)$ $\qquad$ $\log_a MN = \log_a M + \log_a N$

$\quad = \dfrac{1}{2}(1 + \log_6 x)$ $\qquad\qquad$ $\log_a a = 1$

41. $\log_3 \dfrac{1}{\sqrt[3]{3x}} = \log_3 \dfrac{1}{(3x)^{1/3}}$

$\quad = \log_3 1 - \log_3 (3x)^{1/3}$ $\qquad$ $\log_a \dfrac{M}{N} = \log_a M - \log_a N$

$\quad = 0 - \dfrac{1}{3} \log_3 (3x)$ $\qquad$ $\log_a 1 = 0, \; \log_a M^k = k \; \log_a M$

$\quad = -\dfrac{1}{3}(\log_3 3 + \log_3 x)$ $\qquad$ $\log_a MN = \log_a M + \log_a N$

$\quad = -\dfrac{1}{3}(1 + \log_3 x)$ $\qquad\qquad$ $\log_a a = 1$

45. $\log_{10} \dfrac{x}{\sqrt{x+2}} = \log_{10} \dfrac{x}{(x+2)^{1/2}}$

$\quad = \log_{10} x - \log_{10} (x+2)^{1/2}$ $\qquad$ $\log_a \dfrac{M}{N} = \log_a M - \log_a N$

$\quad = \log_{10} x - \dfrac{1}{2} \log_{10} (x+2)$ $\qquad$ $\log_a M^k = k \; \log_a M$

49. $\qquad\qquad 3.62^x = 12.4$

$\quad \log_{10} (3.62)^x = \log_{10} 12.4$

$\quad x \log_{10} 3.62 = \log_{10} 12.4$

$\qquad\qquad x = \dfrac{\log_{10} 12.4}{\log_{10} 3.62} = 1.96$

53. $(36.4)^x = 0.147$

   $\log(36.4)^x = \log 0.147$

   $x \log 36.4 = \log 0.147$

   $x = \dfrac{\log 0.147}{\log 36.4} = -0.533$

Section 6.7

1. $y = \ln 2x$

   $y' = \dfrac{1}{2x} \dfrac{d}{dx}(2x) = \dfrac{1}{2x}(2) = \dfrac{1}{x}$

5. $R = \ln s^2$

   $\dfrac{dR}{ds} = \dfrac{1}{s^2} \dfrac{d}{ds}(s^2) = \dfrac{1}{s^2}(2s) = \dfrac{2}{s}$

   or

   $R = \ln s^2 = 2 \ln s$; thus $\dfrac{dR}{ds} = \dfrac{2}{s}$

9. $y = \log_{10} x^3 = 3 \log_{10} x$; $y' = \dfrac{3}{x} \log_{10} e$

13. $y = x \ln x$. By the product rule,

   $y' = x \dfrac{d}{dx} \ln x + \ln x \cdot 1 = x \dfrac{1}{x} + \ln x = 1 + \ln x$

17. $y = \ln \dfrac{1}{\sqrt{x+2}} = \ln(x+2)^{-1/2}$

   $= -\dfrac{1}{2} \ln(x+2)$ $\log_a M^k = k \log_a M$

   $y' = -\dfrac{1}{2} \dfrac{1}{x+2} = -\dfrac{1}{2(x+2)}$

21. $y = \ln \dfrac{x^2}{x+1} = \ln x^2 - \ln(x+1)$ $\log_a \dfrac{M}{N} = \log_a M - \log_a N$

   $= 2 \ln x - \ln(x+1)$ $\log_a M^k = k \log_a M$

   $y' = 2 \cdot \dfrac{1}{x} - \dfrac{1}{x+1}$

   $= \dfrac{2}{x} \dfrac{x+1}{x+1} - \dfrac{1}{x+1} \dfrac{x}{x} = \dfrac{2x+2-x}{x(x+1)}$

   $y' = \dfrac{x+2}{x(x+1)}$

25. $y = \ln \dfrac{\sqrt{x}}{2-x^2} = \ln \dfrac{x^{1/2}}{2-x^2}$

   $= \ln x^{1/2} - \ln(2-x^2)$ $\log_a \dfrac{M}{N} = \log_a M - \log_a N$

126

$$= \tfrac{1}{2} \ln x - \ln(2 - x^2) \qquad\qquad \log_a M^k = k \log_a M$$

$$y' = \tfrac{1}{2} \cdot \tfrac{1}{x} - \frac{1}{2 - x^2} \frac{d}{dx}(2 - x^2)$$

$$= \frac{1}{2x} - \frac{-2x}{2 - x^2} = \frac{1}{2x} \cdot \frac{2 - x^2}{2 - x^2} + \frac{2x}{2 - x^2} \cdot \frac{2x}{2x}$$

$$= \frac{2 - x^2 + 4x^2}{2x(2 - x^2)} = \frac{2 + 3x^2}{2x(2 - x^2)}$$

29. $y = \ln \dfrac{\sec^2 x}{\sqrt{x + 1}} = \ln \sec^2 x - \ln(x + 1)^{1/2}$

$$= 2 \ln \sec x - \tfrac{1}{2} \ln(x + 1)$$

$$y' = 2 \frac{\sec x \tan x}{\sec x} - \frac{1}{2(x + 1)}$$

$$= 2 \tan x - \frac{1}{2(x + 1)} = \frac{4(x + 1)\tan x - 1}{2(x + 1)}$$

33. $y = (x + 1)^{1/2} \ln x.$ By the product rule,

$$y' = (x + 1)^{1/2} \tfrac{1}{x} + \tfrac{1}{2}(x + 1)^{-1/2} \ln x = \frac{\sqrt{x + 1}}{x} + \frac{\ln x}{2\sqrt{x + 1}}$$

37. $y = \ln[1 + (x^2 - 1)^{1/2}]$

$$y' = \frac{1}{1 + (x^2 - 1)^{1/2}} \frac{d}{dx}[1 + (x^2 - 1)^{1/2}]$$

$$= \frac{1}{1 + \sqrt{x^2 - 1}}\left[\tfrac{1}{2}(x^2 - 1)^{-1/2}(2x)\right] = \frac{x}{(1 + \sqrt{x^2 - 1})\sqrt{x^2 - 1}}$$

$$= \frac{x}{\sqrt{x^2 - 1} + (\sqrt{x^2 - 1})^2} = \frac{x}{x^2 - 1 + \sqrt{x^2 - 1}}$$

41. $y = \ln(\ln x)$

$$y' = \frac{1}{\ln x} \frac{d}{dx}(\ln x) = \frac{1}{\ln x} \cdot \tfrac{1}{x} = \frac{1}{x \ln x}$$

Section 6.8

1. $y = e^{4x}$

$$y' = e^{4x} \frac{d}{dx}(4x) = 4e^{4x}$$

5. $y = 2e^{-t^2}$

$$\frac{dy}{dt} = 2e^{-t^2} \frac{d}{dt}(-t^2) = 2e^{-t^2}(-2t) = -4te^{-t^2}$$

9. $y = 3^{x^3}$

$y' = 3^{x^3}(\ln 3)\frac{d}{dx}(x^3) = 3^{x^3}(\ln 3)(3x^2)$

13. $C = 2re^r$. By the product rule,

$\frac{dC}{dr} = 2r\frac{d}{dr}e^r + e^r\frac{d}{dr}(2r) = 2re^r + 2e^r = 2(r+1)e^r$

17. $y = \sin e^x$

$y' = (\cos e^x)\frac{d}{dx}e^x = e^x \cos e^x$

21. $y = \dfrac{\tan x}{e^{x^2}}$. By the quotient rule,

$y' = \dfrac{e^{x^2}\sec^2 x - (\tan x)e^{x^2}(2x)}{(e^{x^2})^2} = \dfrac{e^{x^2}(\sec^2 x - 2x \tan x)}{(e^{x^2})^2}$

$= \dfrac{\sec^2 x - 2x \tan x}{e^{x^2}}$

25. $y = \ln(\sin e^{2x})$

$y' = \dfrac{1}{\sin e^{2x}}\dfrac{d}{dx}(\sin e^{2x}) = \dfrac{1}{\sin e^{2x}}(\cos e^{2x})\dfrac{d}{dx}e^{2x}$

$= \dfrac{1}{\sin e^{2x}}(\cos e^{2x})e^{2x}(2) = 2e^{2x}\dfrac{\cos e^{2x}}{\sin e^{2x}} = 2e^{2x}\cot e^{2x}$

29. $\dfrac{d}{dx}\sinh x = \dfrac{d}{dx}\left[\dfrac{1}{2}(e^x - e^{-x})\right] = \dfrac{1}{2}[e^x - e^{-x}(-1)]$

$= \dfrac{1}{2}(e^x + e^{-x}) = \cosh x$

$\dfrac{d}{dx}\cosh x = \dfrac{d}{dx}\left[\dfrac{1}{2}(e^x + e^{-x})\right] = \dfrac{1}{2}[e^x + e^{-x}(-1)]$

$= \dfrac{1}{2}(e^x - e^{-x}) = \sinh x$

33. $y = \cosh 2x^3$
$y' = \sinh 2x^3 \cdot 6x^2 = 6x^2 \sinh 2x^3$

37. $y = (\ln x)^x$
$\ln y = \ln(\ln x)^x$
$\qquad = x \ln(\ln x)$ property (6.18)

$\dfrac{1}{y}\dfrac{dy}{dx} = x\dfrac{d}{dx}\ln(\ln x) + \ln(\ln x)\cdot 1 = x\dfrac{1}{\ln x}\dfrac{1}{x} + \ln(\ln x)$

$\qquad = \dfrac{1}{\ln x} + \ln(\ln x)$

$\dfrac{dy}{dx} = y\left[\dfrac{1}{\ln x} + \ln(\ln x)\right]$

Section 6.9

1. $\displaystyle \lim_{x \to -2} \frac{x^2 - 4}{x + 2} = \lim_{x \to -2} \frac{\frac{d}{dx}(x^2 - 4)}{\frac{d}{dx}(x + 2)} = \lim_{x \to -2} \frac{2x}{1} = -4$

5. $\displaystyle \lim_{t \to 3} \frac{t^2 + t - 12}{t - 3} = \lim_{t \to 3} \frac{\frac{d}{dt}(t^2 + t - 12)}{\frac{d}{dt}(t - 3)}$

$\displaystyle = \lim_{t \to 3} \frac{2t + 1}{1} = 7$

9. $\displaystyle \lim_{x \to 0} \frac{\tan 3x}{1 - \cos x} = \lim_{x \to 0} \frac{3 \sec^2 3x}{\sin x} = \infty$ (limit does not exist)

13. $\displaystyle \lim_{x \to 0} \frac{1 - e^x}{2x} = \lim_{x \to 0} \frac{-e^x}{2} = -\frac{1}{2}$

17. $\displaystyle \lim_{x \to 0} \frac{x + \sin 2x}{x - \sin 2x} = \lim_{x \to 0} \frac{1 + 2 \cos 2x}{1 - 2 \cos 2x} = \frac{1 + 2}{1 - 2} = -3$

21. $\displaystyle \lim_{x \to 0^+} (\sin x) \ln \sin x = \lim_{x \to 0^+} \frac{\ln \sin x}{(\sin x)^{-1}}$

$\displaystyle = \lim_{x \to 0^+} \frac{\frac{1}{\sin x} \cos x}{-1(\sin x)^{-2} \cos x} = \lim_{x \to 0^+} \left(-\frac{\frac{1}{\sin x}}{(\sin x)^{-2}} \right)$

$\displaystyle = \lim_{x \to 0^+} (-\sin x) = 0$

Section 6.10

1. $f(x) = xe^{-x}$
    $f'(x) = xe^{-x}(-1) + e^{-x} = e^{-x} - xe^{-x}$
    $f''(x) = -e^{-x} - xe^{-x}(-1) - e^{-x} = xe^{-x} - 2e^{-x}$

Step 1. Critical points:
        $f'(x) = e^{-x} - xe^{-x} = e^{-x}(1 - x) = 0$;
        thus $x = 1$. From $y = xe^{-x}$, the point $(1, 1/e)$ is the critical point.

Step 2. Test of critical point:
        $f''(1) = xe^{-x} - 2e^{-x}\big|_{x=1} = e^{-1} - 2e^{-1} = -\frac{1}{e} < 0$

        Thus $(1, 1/e)$ is a maximum.

Step 3. Concavity and points of inflection:
        $f''(x) = xe^{-x} - 2e^{-x} = 0$

129

$$e^{-x}(x - 2) = 0 \; ;$$

thus $x = 2$. From $y = xe^{-x}$, the point is $(2, 2/e^2)$.

If $x < 2$, $y'' < 0$ (concave down)

If $x > 2$, $y'' > 0$ (concave up)

Step 4. Other:

$$\lim_{x \to +\infty} xe^{-x} = \lim_{x \to +\infty} \frac{x}{e^x} = \lim_{x \to +\infty} \frac{1}{e^x} = 0$$

by L'Hospital's rule. So the positive x-axis is an asymptote.

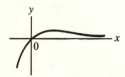

5. $f(x) = \cosh x = \frac{1}{2}(e^x + e^{-x})$

$f'(x) = \sinh x = \frac{1}{2}(e^x - e^{-x})$

$f''(x) = \cosh x = \frac{1}{2}(e^x + e^{-x})$

Step 1. Critical points

$$f'(x) = \frac{1}{2}(e^x - e^{-x}) = 0 \text{ when } x = 0$$

Step 2. Test of critical point

$$f''(0) = \frac{1}{2}(1 + 1) = 1 > 0$$

The point $(0,1)$ is a minimum.

Step 3. Concavity

$$f''(x) = \frac{1}{2}(e^x + e^{-x}) > 0$$

The graph is concave up. (There are no inflection points.)

9. Recall that $v = L(\frac{di}{dt})$. Thus

$v = 0.0030 \frac{d}{dt}[3.0 \sin 200t + 2.0 \cos 200t]$

$= (0.0030)[(3.0)(200) \cos 200t - (2.0)(200)\sin 200t]_{t=0}$

$= (0.0030)[(3.0)(200)(1) - 0] = 1.8 \text{ V}$

13. $N = 6.0e^{-0.25t}$

$\frac{dN}{dt} = 6.0e^{-0.25t}(-0.25)|_{t=8.5}$

$= 6.0e^{(-0.25)(8.5)}(-0.25) = -0.18$ g/min

17. $S = kR^2 \ln \frac{1}{R} = kR^2(\ln 1 - \ln R) = -kR^2 \ln R$, since $\ln 1 = 0$.
By the product rule

$\frac{dS}{dR} = -kR^2 \frac{1}{R} - 2kR \ln R = -kR - 2kR \ln R$

$= -kR(1 + 2 \ln R) = 0, \quad R \neq 0$

$2 \ln R = -1$

$\ln R = -\frac{1}{2}$

By the definition of logarithm,

$e^{-1/2} = R \quad$ or $\quad R = \frac{1}{\sqrt{e}}$

21. In a problem on related rates, we label all quantities that vary by
letters, as shown in the figure.

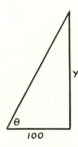

Given $\frac{dy}{dt} = 6.0$, find $\frac{d\theta}{dt}$ when $y = 150$.

From the figure:

$\tan \theta = \frac{y}{100}$

$\theta = \text{Arctan} \frac{y}{100}$

$\frac{d\theta}{dt} = \frac{1}{1 + (y/100)^2}\left(\frac{1}{100}\right)\frac{dy}{dt}$

$= \frac{1}{1 + (y/100)^2}\left(\frac{1}{100}\right)(6.0)|_{y=150}$

$= 0.018$ rad/s

25.

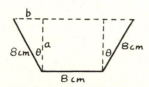

To find an expression for the area, we need to determine a and b in
the figure. Observe that

$\cos \theta = \frac{a}{8} \quad$ and $\quad \sin \theta = \frac{b}{8}$

Hence

$a = 8 \cos \theta$ and $b = 8 \sin \theta$

Area of each triangle: $\frac{1}{2}ab = \frac{1}{2}(64) \sin \theta \cos \theta$

Area of rectangle: $8a = 64 \cos \theta$

Total area $A = 64 \cos \theta + 64 \sin \theta \cos \theta$

$\qquad = 64(\cos \theta + \sin \theta \cos \theta)$

$\dfrac{dA}{d\theta} = 64[-\sin \theta + \sin \theta(-\sin \theta) + \cos \theta(\cos \theta)]$

$\qquad = 64(-\sin \theta - \sin^2 \theta + \cos^2 \theta)$

$\qquad = 64(-\sin \theta - \sin^2 \theta + 1 - \sin^2 \theta)$ $\qquad\qquad \cos^2 \theta = 1 - \sin^2 \theta$

$\qquad = -64(2 \sin^2 \theta + \sin \theta - 1)$

$\qquad = -64(2 \sin \theta - 1)(\sin \theta + 1) = 0$ $\qquad\qquad$ factoring

$2 \sin \theta = 1$

$\quad \sin \theta = \frac{1}{2}$ and $\theta = \frac{\pi}{6}$

29. $R = k \cos \theta \sin(\theta - \alpha)$, where $k = \dfrac{2v_0^2}{g \cos^2 \alpha}$. By the product rule,

$\dfrac{dR}{d\theta} = k[\cos \theta \cos(\theta - \alpha) - \sin \theta \sin(\theta - \alpha)]$

From the identity $\cos(A + B) = \cos A \cos B - \sin A \sin B$,

$\dfrac{dR}{d\theta} = k \cos[\theta + (\theta - \alpha)] = k \cos(2\theta - \alpha) = 0$

Thus

$2\theta - \alpha = \frac{\pi}{2}$

and

$\theta = \frac{\pi}{4} + \frac{\alpha}{2}$

Section 6.11

1. $f(x) = \cos x + x$

$f'(x) = -\sin x + 1$

By Newton's method

$x_1 = x_0 - \dfrac{\cos x_0 + x_0}{-\sin x_0 + 1} = x_0 + \dfrac{\cos x_0 + x_0}{\sin x_0 - 1}$

Since the graph of $y = -x$ intersects the graph of $y = \cos x$ between $-\frac{\pi}{2}$ and 0, we choose $x_0 = -1$. Thus

$x_1 = -1 + \dfrac{\cos(-1) + (-1)}{\sin(-1) - 1}$

Setting the calculator in the radian mode, we get

$x_1 = -0.75$

It is best to store this value in the memory when computing the next approximation x_2:

$$x_2 = -0.75 + \frac{\cos(-0.75) + (-0.75)}{\sin(-0.75) - 1} = -0.739$$

A possible sequence is

0.75 $\boxed{+/-}$ $\boxed{\text{STO}}$ $\boxed{\text{COS}}$ $\boxed{+}$ $\boxed{\text{RCL}}$ $\boxed{=}$ $\boxed{\div}$ $\boxed{(}$ $\boxed{\text{RCL}}$ $\boxed{\text{SIN}}$
$\boxed{=}$ 1 $\boxed{)}$ $\boxed{+}$ $\boxed{\text{RCL}}$ $\boxed{=}$ $\boxed{\text{STO}}$ $\rightarrow$ -0.7391111

Having stored the last value, we proceed as before:

$$x_3 = -0.739 + \frac{\cos(-0.739) + (-0.739)}{\sin(-0.739) - 1} = -0.7390851$$

From this point on, these digits no longer change, so that $x = -0.7390851$ is the root of the equation to 7 decimal places.

5. $f(x) = \sin x - x^2$

$f'(x) = \cos x - 2x$

By Newton's method

$$x_1 = x_0 - \frac{\sin x_0 - x_0^2}{\cos x_0 - 2x_0} = x_0 + \frac{\sin x_0 - x_0^2}{2x_0 - \cos x_0}$$

The parabola $y = x^2$ intersects the curve $y = \sin x$ between 0 and $\frac{\pi}{2}$. So we choose $x_0 = 1$:

$$x_1 = 1 + \frac{\sin 1 - 1}{2 - \cos 1} = 0.89$$

Store 0.89 in the memory and evaluate

$$x_2 = 0.89 + \frac{\sin(0.89) - (0.89)^2}{2(0.89) - \cos(0.89)} = 0.8769$$

Store 0.8769 in the memory and evaluate

$$x_3 = 0.8769 + \frac{\sin(0.8769) - (0.8769)^2}{2(0.8769) - \cos(0.8769)} = 0.8767262$$

From this point on, these digits no longer change, so that $x = 0.8767262$ is the root of the equation to 7 decimal places.

9. $\tan x + x = 2$; since the line $y = 2 - x$ intersects $y = \tan x$ between 0 and $\frac{\pi}{2}$, we choose $x_0 = 1$, a convenient value.

Let $f(x) = \tan x + x - 2$; then $f'(x) = \sec^2 x + 1$.

$$x_1 = x_0 - \frac{\tan x_0 + x_0 - 2}{\sec^2 x_0 + 1} = 1 - \frac{\tan 1 + 1 - 2}{\sec^2 1 + 1}$$

$$= 0.874$$

$$x_2 = 0.874 - \frac{\tan(0.874) + 0.874 - 2}{\sec^2(0.874) + 1}$$

$$= 0.85387$$

$$x_3 = 0.85387 - \frac{\tan(0.85387) + 0.85387 - 2}{\sec^2(0.85387) + 1}$$

$$= 0.853530$$

13. A graphing calculator will quickly locate the roots near -1, between 0 and 1, and near 6. Suppose we find the middle root by letting $x_0 = 1$:

$$f(x) = x^3 - 6x^2 - 4x + 6$$

$$f'(x) = 3x^2 - 12x - 4$$

$$x_1 = x_0 - \frac{x_0^3 - 6x_0^2 - 4x_0 + 6}{3x_0^2 - 12x_0 - 4} = 1 - \frac{1^3 - 6 \cdot 1^2 - 4 \cdot 1 + 6}{3 \cdot 1^2 - 12 \cdot 1 - 4}$$

$$= 0.7692$$

$$x_2 = 0.7692 - \frac{(0.7692)^3 - 6(0.7692)^2 - 4(0.7692) + 6}{3(0.7692)^2 - 12(0.7692) - 4}$$

$$= 0.754211$$

$$x_3 = 0.754211 - \frac{(0.754211)^3 - 6(0.754211)^2 - 4(0.754211) + 6}{3(0.754211)^2 - 12(0.754211) - 4}$$

$$= 0.754138$$

REVIEW EXERCISES FOR CHAPTER 6

1. In the interval $\left[-\frac{\pi}{2}, \frac{\pi}{2}\right]$, $\sin\theta = -1$ only for $\theta = -\frac{\pi}{2}$.

5. Let $\theta = \text{Arctan}\,\frac{3}{1}$. Then the length of the hypotenuse is $\sqrt{10}$. Thus

$$\cos(\text{Arctan } 3) = \cos\theta$$

$$= \frac{1}{\sqrt{10}}$$

$$= \frac{\sqrt{10}}{10}$$

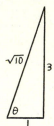

9. $y = x^2 \tan 3x$. By the product rule,

$$y' = x^2 \frac{d}{dx} \tan 3x + \tan 3x \frac{d}{dx}(x^2)$$

$$= x^2 \sec^2 3x \cdot 3 + \tan 3x (2x)$$

$$= 3x^2 \sec^2 3x + 2x \tan 3x$$

13. $y = \ln \dfrac{1}{\sqrt{x+3}} = \ln(x+3)^{-1/2}$

$$= -\frac{1}{2} \ln(x+3) \qquad\qquad \log_a M^k = k \log_a M$$

$$y' = -\frac{1}{2} \frac{1}{x+3} = -\frac{1}{2(x+3)}$$

17. $y = \sqrt{\ln\ 2x} = (\ln\ 2x)^{1/2}$. By the power rule,

$y' = \frac{1}{2}(\ln\ 2x)^{-1/2}\ \frac{d}{dx}\ \ln\ 2x = \frac{1}{2}(\ln\ 2x)^{-1/2}\ \frac{1}{2x}(2)$

$= \dfrac{1}{2x\sqrt{\ln\ 2x}}$

21. $y = e^{2x} \cot x$. By the product rule,

$y' = e^{2x}(-\csc^2 x) + (\cot x)e^{2x}(2)$

$= e^{2x}(2\ \cot x - \csc^2 x)$

25. $y = (\cot x)^x$

$\ln y = \ln(\cot x)^x$

$\qquad = x\ \ln\ \cot x$ by property (6.18)

$\dfrac{1}{y}\ \dfrac{dy}{dx} = x\ \dfrac{d}{dx}\ \ln\ \cot x + \ln\ \cot x \cdot 1$

$\qquad\qquad = x\ \dfrac{1}{\cot x}(-\csc^2 x) + \ln\ \cot x$

Note that

$\dfrac{1}{\cot x}\ \csc^2 x = \dfrac{\sin x}{\cos x}\ \dfrac{1}{\sin^2 x} = \dfrac{1}{\cos x\ \sin x} = \sec x\ \csc x$

Thus

$\dfrac{1}{y}\ \dfrac{dy}{dx} = -x\ \sec x\ \csc x + \ln\ \cot x$

$\dfrac{dy}{dx} = y(\ln\ \cot x - x\ \csc x\ \sec x)$

29. $\displaystyle\lim_{x\to 0}\ \dfrac{\sin\ 4x}{x} = \lim_{x\to 0}\ \dfrac{4\ \cos\ 4x}{1} = 4$ by L'Hospital's rule

33. $\displaystyle\lim_{x\to 0^+}\ \dfrac{\sin x - x}{x\ \sin x} = \lim_{x\to 0^+}\ \dfrac{\cos x - 1}{x\ \cos x + \sin x}$

which tends to the indeterminate form $\frac{0}{0}$. Applying the rule again, we get:

$\displaystyle\lim_{x\to 0^+}\ \dfrac{-\sin x}{-x\ \sin x + \cos x + \cos x} = \dfrac{0}{0 + 1 + 1} = 0$

37. (a) From $P = Se^{rt}$, we get

        $P = 5000e^{(0.10)(15)} = \$22,408.45$

   (b) $\dfrac{dP}{dt} = Se^{rt}\ \dfrac{d}{dt}(rt) = Se^{rt}r = Pr = rP$

      (P grows at a rate that is proportional to the amount present.)

41.

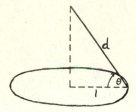

Since d = sec θ and $I = k \dfrac{\sin \theta}{d^2}$, we get

$$I = \frac{k \sin \theta}{\sec^2 \theta} = k \sin \theta \cos^2 \theta$$

Then

$$\frac{dI}{d\theta} = k[\sin \theta (2 \cos \theta)(-\sin \theta) + \cos^2 \theta \cos \theta]$$
$$= k \cos \theta (-2 \sin^2 \theta + \cos^2 \theta)$$
$$= k \cos \theta (-2 \sin^2 \theta + 1 - \sin^2 \theta) \qquad \cos^2 \theta = 1 - \sin^2 \theta$$
$$= k \cos \theta (1 - 3 \sin^2 \theta) = 0$$

$$\sin^2 \theta = \frac{1}{3}$$

$$\sin \theta = \frac{1}{\sqrt{3}}$$

Hence

$$\theta = \text{Arcsin } \frac{1}{\sqrt{3}} \approx 35.3°$$

CHAPTER 7   INTEGRATION TECHNIQUES

<u>Section 7.1</u>

1. $\int x\sqrt{x^2 + 1}\,dx = \int (x^2 + 1)^{1/2}x\,dx$ Let $u = x^2 + 1$; then $du = 2x\,dx$

$\frac{1}{2}\int (x^2 + 1)^{1/2}(2x)\,dx = \frac{1}{2}\int u^{1/2}\,du + C$

$= \frac{1}{2}\,\frac{2}{3}u^{3/2} + C = \frac{1}{3}(x^2 + 1)^{3/2} + C$

5. $\int \sin^2 x\,\cos x\,dx$ Let $u = \sin x$; then $du = \cos x\,dx$

$\int u^2\,du = \frac{1}{3}u^3 + C = \frac{1}{3}\sin^3 x + C$

9. $\int (1 + \tan 3t)^3 \sec^2 3t\,dt$ $u = 1 + \tan 3t$

$du = 3\sec^2 3t\,dt$

$= \frac{1}{3}\int (1 + \tan 3t)^3\,(3\sec^2 3t\,dt)$

$= \frac{1}{3}\int u^3\,du = \frac{1}{3}\,\frac{u^4}{4} + C = \frac{1}{12}(1 + \tan 3t)^4 + C$

13. $\int \frac{\ln x\,dx}{x} = \int \ln x\left(\frac{1}{x}dx\right)$ Let $u = \ln x$; then $du = \frac{1}{x}\,dx$

$\int u\,du = \frac{1}{2}u^2 + C = \frac{1}{2}\ln^2 x + C$

17. $\int \frac{\text{Arccos } x\,dx}{\sqrt{1 - x^2}}$ Let $u = \text{Arccos } x$; then $du = -\frac{dx}{\sqrt{1 - x^2}}$

$-\int \text{Arccos } x\left(-\frac{dx}{\sqrt{1 - x^2}}\right) = -\int u\,du = -\frac{1}{2}u^2 + C = -\frac{1}{2}\left(\text{Arccos } x\right)^2 + C$

21. $\int_1^e \frac{\sqrt{\ln x}}{x}dx = \int (\ln x)^{1/2}\left(\frac{1}{x}dx\right)$ Let $u = \ln x$; then $du = \frac{1}{x}dx$

$\int_1^e (\ln x)^{1/2}\,\frac{1}{x}dx = \frac{2}{3}(\ln x)^{3/2}\Big|_1^e$

$= \frac{2}{3}\left[(\ln e)^{3/2} - (\ln 1)^{3/2}\right] = \frac{2}{3}(1 - 0) = \frac{2}{3}$

(since $\ln e = 1$ and $\ln 1 = 0$)

25. $\int \sqrt{\tan x}\ \sec^2 x\, dx$ Let $u = \tan x$; then $du = \sec^2 x\, dx$

$\int u^{1/2} du = \frac{2}{3} u^{3/2} + C = \frac{2}{3}(\tan x)^{3/2} + C$

29. $\int \frac{(1 + \ln x)^2}{x} dx$ Let $u = 1 + \ln x$; then $du = \frac{1}{x} dx$

$\int u^2 du = \frac{1}{3} u^3 + C = \frac{1}{3}(1 + \ln x)^3 + C$

Section 7.2

1. $\int \frac{dx}{x - 1}$ $u = x - 1$; $du = dx$

$\int \frac{dx}{x - 1} = \int \frac{du}{u} = \ln|u| + C = \ln|x - 1| + C$

5. $\int \frac{ds}{1 - 3s}$ $u = 1 - 3s$; $du = -3ds$

$-\frac{1}{3} \int \frac{-3ds}{1 - 3s} = -\frac{1}{3} \int \frac{du}{u} = -\frac{1}{3}\ln|u| + C$

$\qquad\qquad = -\frac{1}{3}\ln|1 - 3s| + C$

9. $\int e^{-x} dx$ $u = -x$; $du = -dx$

$-\int e^{-x}(-dx) = -\int e^u du = -e^u + C$

$\qquad\qquad = -e^{-x} + C$

13. $\int e^{4x} dx$ $u = 4x$; $du = 4dx$

$\frac{1}{4} \int e^{4x}(4dx) = \frac{1}{4} \int e^u du = \frac{1}{4} e^u + C$

$\qquad\qquad = \frac{1}{4} e^{4x} + C$

17. $\int e^{\sin R} \cos R\, dR$ $u = \sin R$; $du = \cos R\, dR$

$\int e^u du = e^u + C = e^{\sin R} + C$

21. $\int \frac{e^{\text{Arctan} x}}{1 + x^2} dx = \int e^{\text{Arctan} x}\left(\frac{1}{1 + x^2} dx\right)$ $u = \text{Arctan } x$; $du = \frac{1}{1 + x^2} dx$

$\int e^u du = e^u + C = e^{\text{Arctan} x} + C$

25. $\displaystyle\int \frac{e^x + 1}{e^x}dx = \int\left(\frac{e^x}{e^x} + \frac{1}{e^x}\right)dx$

$$= \int(1 + e^{-x})dx \qquad\qquad u = -x;\ du = -dx$$

$$-\int(1 + e^{-x})(-dx) = -\int(1 + e^u)du = -(u + e^u) + C$$

$$= -u - e^u + C = x - e^{-x} + C$$

29. $\displaystyle\int \frac{2x}{(1 + x^2)^2}dx = \int(1 + x^2)^{-2}(2x)dx \qquad u = 1 + x^2;\ du = 2x\,dx$

$$\int u^{-2}du = \frac{u^{-1}}{-1} + C = -\frac{1}{u} + C = -\frac{1}{1 + x^2} + C$$

33. $\displaystyle\int \frac{x + 1}{x + 2}dx = \int \frac{(x + 2) - 1}{x + 2}dx$

$$= \int\left(1 - \frac{1}{x + 2}\right)dx$$

which is equivalent to long division.

$$\qquad\qquad\qquad u = x + 2;\ du = dx$$

$$\int 1\,dx - \int \frac{dx}{x + 2} = \int dx - \int \frac{du}{u}$$

$$= x - \ln|u| + C = x - \ln|x + 2| + C$$

37. $\displaystyle\int e^{\sin^2 x}\sin 2x\,dx = \int e^{\sin^2 x}(2\sin x \cos x)dx$

by the double-angle formula.

$$\qquad\qquad\qquad u = \sin^2 x;\ du = 2(\sin x)(\cos x\,dx)$$

$$\int e^u du = e^u + C = e^{\sin^2 x} + C$$

41. $\displaystyle\int 5^{3x}dx \qquad\qquad\qquad u = 3x;\ du = 3\,dx$

$$\frac{1}{3}\int 5^{3x}(3\,dx) = \frac{1}{3}\int 5^u du = \frac{1}{3}(5^u)\log_5 e = \frac{1}{3}(5^{3x})\log_5 e + C$$

45.

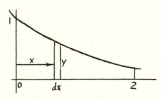

Moment of inertia (with respect to y-axis) of typical element:
$\rho x^2 y\ dx$

$$I_y = \rho \int_0^2 x^2 \frac{1}{1 + x^3}dx \qquad\qquad \text{Let } u = 1 + x^3; \text{ then } du = 3x^2 dx$$

139

$$I_y = \frac{\rho}{3} \int_0^2 \frac{3x^2\,dx}{1 + x^3} = \frac{\rho}{3} \ln|1 + x^3|\,\Big|_0^2 = \frac{\rho}{3}[\ln 9 - \ln 1]$$

$$= \frac{\rho}{3}[\ln 3^2 - 0] = \frac{\rho}{3}(2 \ln 3) = \frac{2}{3}\rho \ln 3$$

49.

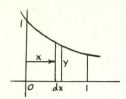

Volume of shell: $2\pi \cdot$ (radius) $\cdot$ (height) $\cdot$ (thickness)
$$= 2\pi x y\,dx$$

$$V = 2\pi \int_0^1 x e^{-x^2}\,dx \qquad\qquad \text{Let } u = -x^2; \text{ then } du = -2x\,dx$$

$$V = \frac{2\pi}{-2} \int_0^1 e^{-x^2}(-2x)\,dx = -\pi e^{-x^2}\Big|_0^1 = -\pi e^{-1} + \pi \cdot 1$$

$$= \pi(-e^{-1} + 1) = \pi\left(1 - \frac{1}{e}\right)$$

53. Let $u = 10 - kv$; then $du = -k\,dv$. Multiplying both sides by $-k$, we get

$$\frac{-k\,dv}{10 - kv} = -k\,dt$$

$\ln|10 - kv| = -kt + C$ \qquad\qquad Integrating

Since $10 - kv > 0$, $\ln|10 - kv| = \ln(10 - kv)$. Thus

$\ln(10 - kv) = -kt + C$

If $t = 0$, $v = 0$. Hence $\ln 10 = 0 + C$ and

$\ln(10 - kv) = -kt + \ln 10$

$\ln(10 - kv) - \ln 10 = -kt$

$\ln \dfrac{10 - kv}{10} = -kt$ \qquad\qquad $\log_a M - \log_a N = \log_a \dfrac{M}{N}$

By definition of logarithm,

$$e^{-kt} = \frac{10 - kv}{10} \qquad\qquad kv = 10(1 - e^{-kt})$$

$$10e^{-kt} = 10 - kv \qquad\qquad v = \frac{10}{k}(1 - e^{-kt}) \text{ m/s}$$

Section 7.3

1. $\displaystyle\int \sec^2 2x\,dx$ \qquad\qquad $u = 2x;\ du = 2\,dx$

$$\frac{1}{2}\int \sec^2 2x(2\,dx) = \frac{1}{2}\int \sec^2 u\,du = \frac{1}{2}\tan u + C = \frac{1}{2}\tan 2x + C$$

5. $\displaystyle\int \csc 4x \cot 4x\, dx$ $\qquad\qquad u = 4x;\ du = 4dx$

$\displaystyle\frac{1}{4}\int \csc 4x \cot 4x(4dx) = \frac{1}{4}\int \csc u \cot u\, du = \frac{1}{4}(-\csc u) + C$

$$= -\frac{1}{4}\csc 4x + C$$

9. $\displaystyle\int \cos 2t\, dt$ $\qquad\qquad u = 2t;\ du = 2dt$

$\displaystyle\frac{1}{2}\int \cos 2t\,(2dt) = \frac{1}{2}\int \cos u\, du = \frac{1}{2}\sin u + C = \frac{1}{2}\sin 2t + C$

13. $\displaystyle\int x \sin x^2\, dx$ $\qquad\qquad u = x^2;\ du = 2x\, dx$

$\displaystyle\frac{1}{2}\int \sin x^2(2x)dx = \frac{1}{2}\int \sin u\, du = -\frac{1}{2}\cos u + C$

$$= -\frac{1}{2}\cos x^2 + C$$

17. $\displaystyle\int \frac{\cos \sqrt{x}}{\sqrt{x}}\, dx = \int \frac{\cos x^{1/2}}{x^{1/2}}dx$ $\qquad$ Let $u = x^{1/2}$; then
$$du = \frac{1}{2}x^{-1/2}dx = \frac{dx}{2x^{1/2}}$$

$\displaystyle 2\int \cos x^{1/2}\,\frac{dx}{2x^{1/2}} = 2\int \cos u\, du = 2\sin u + C = 2\sin \sqrt{x} + C$

21. $\displaystyle\int \csc e^{3x}(e^{3x})dx$ $\qquad\qquad u = e^{3x};\ du = 3e^{3x}dx$

$\displaystyle\frac{1}{3}\int \csc e^{3x}(3e^{3x})dx = \frac{1}{3}\int \csc u\, du = \frac{1}{3}\ln|\csc u - \cot u| + C$

$$= \frac{1}{3}\ln|\csc e^{3x} - \cot e^{3x}| + C$$

25. $\displaystyle\int \frac{\sec^2 x}{\tan^2 x}\, dx = \int \sec^2 x \cot^2 x\, dx = \int \frac{1}{\cos^2 x}\,\frac{\cos^2 x}{\sin^2 x}\, dx$

$$= \int \frac{dx}{\sin^2 x} = \int \csc^2 x\, dx = -\cot x + C$$

29. $\displaystyle\int_0^{\sqrt{\pi/2}} \omega \cos \omega^2\, d\omega$ $\qquad\qquad$ Let $u = \omega^2$; then $du = 2\omega\, d\omega$

Lower limit: if $\omega = 0$, $u = 0$
Upper limit: if $\omega = \sqrt{\frac{\pi}{2}}$, $u = \omega^2 = \frac{\pi}{2}$

$\displaystyle\int_0^{\sqrt{\pi/2}} \omega \cos \omega^2\, d\omega = \frac{1}{2}\int_0^{\sqrt{\pi/2}} (\cos \omega^2)(2\omega\, d\omega)$

$\displaystyle = \frac{1}{2}\int_0^{\pi/2} \cos u\, du = \frac{1}{2}\sin u\Big|_0^{\pi/2}$

$\displaystyle = \frac{1}{2}\sin \frac{\pi}{2} - \frac{1}{2}\sin 0 = \frac{1}{2}\cdot 1 - \frac{1}{2}\cdot 0 = \frac{1}{2}$

33. Recall that $v = \frac{q}{C} = \frac{1}{C} \int i\,dt$. Thus

$v = \dfrac{1}{100 \times 10^{-6}} \int 2.00 \cos 100t\,dt$ Let $u = 100t$; then $du = 100\,dt$

$\qquad = \dfrac{2.00}{10^{-4}} \int \cos 100t\,dt = \dfrac{2.00}{10^{-4}} \dfrac{1}{100} \int \cos 100t\,(100)\,dt$

$\qquad = \dfrac{2.00}{10^{-2}} \int \cos u\,du = 200 \sin u + C = 200 \sin 100t + C$

If $t = 0$, $v = 0$.   Thus $C = 0$.

$v = 200 \sin 100t\big|_{t=0.200} = 183$ V,

using a calculator set in radian mode.

37. Moment (with respect to y-axis) of typical element: $xy\,dx$

$M_y = \displaystyle\int_0^{\sqrt{\pi}/2} x \tan x^2\,dx$ Let $u = x^2$; then $du = 2x\,dx$

$M_y = \dfrac{1}{2} \displaystyle\int_0^{\sqrt{\pi}/2} \tan x^2\,(2x)\,dx = \dfrac{1}{2} \ln|\sec x^2|\Big|_0^{\sqrt{\pi}/2}$

$\qquad = \dfrac{1}{2}(\ln \sec \dfrac{\pi}{4} - \ln \sec 0) = \dfrac{1}{2}(\ln \sqrt{2} - \ln 1)$

$\qquad = \dfrac{1}{2}(\ln 2^{1/2} - 0) = \dfrac{1}{2} \dfrac{1}{2} \ln 2 = \dfrac{1}{4} \ln 2$

41. $\displaystyle\int_1^2 \dfrac{\sin x}{x}\,dx$; $h = \dfrac{2 - 1}{8} = \dfrac{1}{8} = 0.125$

$x_0 = 1$, $x_1 = 1.125$, $x_2 = 1.125 + 0.125 = 1.25$,

$x_3 = 1.25 + 0.125 = 1.375$,

The function values are listed next:

$f(1) = \dfrac{\sin 1}{1} = 0.8415$

$f(1.125) = \dfrac{\sin 1.125}{1.125} = 0.8020$

$f(1.25) = \dfrac{\sin 1.25}{1.25} = 0.7592$

$f(1.375) = 0.7134$

$f(1.5) = 0.6650$

$f(1.625) = 0.6145$

$f(1.75) = 0.5623$

$f(1.875) = 0.5088$

$f(2) = 0.4546$

$\displaystyle\int_1^2 \dfrac{\sin x}{x}\,dx \approx 0.125\Big[\dfrac{1}{2}(0.8415) + 0.8020 + 0.7592 + \ldots + \dfrac{1}{2}(0.4546)\Big]$

$\qquad\qquad = 0.66$

1. $\int \sin^2 2x \cos^3 2x\,dx$

TYPE 1, m odd, identity (7.14)

$\int \sin^2 2x \cos^2 2x (\cos 2x\,dx)$

$= \int \sin^2 2x (1 - \sin^2 2x)(\cos 2x\,dx)$ $u = \sin 2x;\ du = 2\cos 2x\,dx$

$= \tfrac{1}{2} \int \sin^2 2x (1 - \sin^2 2x)(2\cos 2x\,dx)$

$= \tfrac{1}{2} \int u^2 (1 - u^2)\,du = \tfrac{1}{2} \int (u^2 - u^4)\,du = \tfrac{1}{2}\left(\tfrac{1}{3}u^3 - \tfrac{1}{5}u^5\right) + C$

$= \tfrac{1}{6} \sin^3 2x - \tfrac{1}{10} \sin^5 2x + C$

5. $\int \sin^3 x \cos^4 x\,dx$

TYPE 1, n odd, identity (7.14)

$\int \sin^2 x \cos^4 x (\sin x\,dx)$

$= \int (1 - \cos^2 x)\cos^4 x (\sin x\,dx)$ $u = \cos x;\ du = -\sin x\,dx$

$= - \int (1 - \cos^2 x)\cos^4 x (-\sin x\,dx)$

$= - \int (1 - u^2)u^4\,du = - \int (u^4 - u^6)\,du = -\tfrac{1}{5}u^5 + \tfrac{1}{7}u^7 + C$

$= \tfrac{1}{7} \cos^7 x - \tfrac{1}{5} \cos^5 x + C$

9. $\int \cos^2 4x\,dx$

TYPE 1, even powers, identity (7.17)

$\int \cos^2 4x\,dx = \tfrac{1}{2} \int (1 + \cos 8x)\,dx$

$= \tfrac{1}{2} \int 1\,dx + \tfrac{1}{2} \int \cos 8x\,dx$ $u = 8x;\ du = 8\,dx$

$= \tfrac{1}{2} \int dx + \tfrac{1}{2}\,\tfrac{1}{8} \int \cos 8x(8\,dx)$

$= \tfrac{1}{2} \int dx + \tfrac{1}{16} \int \cos u\,du = \tfrac{1}{2}x + \tfrac{1}{16} \sin 8x + C$

13. $\int \sin^3 2t \cos^2 2t\,dt$

TYPE 1, n odd

$\int \sin^2 2t \cos^2 2t (\sin 2t\,dt)$

$$= \int (1 - \cos^2 2t)\cos^2 2t(\sin 2t\,dt) \qquad u = \cos 2t; \; du = -2 \sin 2t\,dt$$

$$= -\tfrac{1}{2} \int (1 - \cos^2 2t)\cos^2 2t(-2 \sin 2t\,dt)$$

$$= -\tfrac{1}{2} \int (1 - u^2)u^2\,du = -\tfrac{1}{2} \int (u^2 - u^4)\,du$$

$$= -\tfrac{1}{2}\left(\tfrac{u^3}{3} - \tfrac{u^5}{5}\right) + C = \tfrac{1}{10}u^5 - \tfrac{1}{6}u^3 + C$$

$$= \tfrac{1}{10}\cos^5 2t - \tfrac{1}{6}\cos^3 2t + C$$

17. $\displaystyle\int \tan^3 x\,dx$

TYPE 2, m = 0, identity (7.15)

$$\int \tan^3 x\,dx = \int \tan x \, \tan^2 x\,dx$$

$$= \int \tan x(\sec^2 x - 1)\,dx$$

$$= \int \tan x \, \sec^2 x\,dx - \int \tan x\,dx$$

Let u = tan x; then du = $\sec^2 x\,dx$

$$\int u\,du - \int \tan x\,dx = \tfrac{1}{2}u^2 - (-\ln|\cos x|) + C$$

$$= \tfrac{1}{2} \tan^2 x + \ln|\cos x| + C$$

using formula (7.10)

21. $\displaystyle\int \tan y \, \sec^3 y\,dy$

TYPE 2, n odd

$$\int \tan y \, \sec^3 y\,dy = \int \sec^2 y(\sec y \, \tan y)\,dy$$

Let u = sec y; then du = sec y tan y dy

$$\int u^2\,du = \tfrac{1}{3}u^3 + C = \tfrac{1}{3}\sec^3 y + C$$

25. $\displaystyle\int \cot^6 2x \, \csc^4 2x \, dx$

TYPE 3, m even, identity (7.16)

The technique is the same as for TYPE 2 with m even: set aside $\csc^2 2x$ for du, and change the remaining cosecants to cotangents.

$$\int \cot^6 2x \, \csc^4 2x\,dx = \int \cot^6 2x \, \csc^2 2x(\csc^2 2x)\,dx$$

$$= \int \cot^6 2x(1 + \cot^2 2x)(\csc^2 2x \, dx)$$

$$u = \cot 2x; \quad du = -2 \csc^2 2x \, dx$$

$$-\frac{1}{2}\int \cot^6 2x(1 + \cot^2 2x)(-2 \csc^2 2x \, dx)$$

$$= -\frac{1}{2}\int u^6(1 + u^2)\,du = -\frac{1}{2}\int(u^6 + u^8)\,du = -\frac{1}{2}\left(\frac{1}{7}u^7 + \frac{1}{9}u^9\right) + C$$

$$= -\frac{1}{14}\cot^7 2x - \frac{1}{18}\cot^9 2x + C$$

29. $\displaystyle\int_0^{\pi/4}(\tan x)^{1/2}\sec^4 x \, dx$

TYPE 2, m even, identity (7.15)

$$\int(\tan x)^{1/2}\sec^2 x(\sec^2 x \, dx)$$

$$= \int(\tan x)^{1/2}(1 + \tan^2 x)(\sec^2 x \, dx) \qquad u = \tan x; \quad du = \sec^2 x \, dx$$

$$\int u^{1/2}(1 + u^2)\,du = \int(u^{1/2} + u^{5/2})\,du = \frac{2}{3}u^{3/2} + \frac{2}{7}u^{7/2} + C$$

$$\int_0^{\pi/4}(\tan x)^{1/2}\sec^4 x \, dx = \frac{2}{3}(\tan x)^{3/2} + \frac{2}{7}(\tan x)^{7/2}\Big|_0^{\pi/4}$$

$$= \frac{2}{3}\left(\tan \frac{\pi}{4}\right)^{3/2} + \frac{2}{7}\left(\tan \frac{\pi}{4}\right)^{7/2} - 0$$

$$= \frac{2}{3} + \frac{2}{7} = \frac{20}{21}$$

33. $\displaystyle\int_0^{\pi/4}\frac{\sec^2 x}{1 + \tan x}\,dx$

Let $u = 1 + \tan x$; then $du = \sec^2 x \, dx$

Lower limit: if $x = 0$, $u = 1 + \tan 0 = 1$

Upper limit: if $x = \frac{\pi}{4}$, $u = 1 + \tan \frac{\pi}{4} = 1 + 1 = 2$

$$\int_0^{\pi/4}\frac{\sec^2 x}{1 + \tan x}\,dx = \int_1^2 \frac{du}{u} = \ln|u|\Big\|_1^2 = \ln 2 - \ln 1$$

$$= \ln 2 - 0 = \ln 2$$

37. Volume of disk: $\pi y^2 \, dx$

$$V = \pi \int_0^\pi \sin^2 x \, dx$$

TYPE 1, even powers, identity (7.18)

$$V = \frac{\pi}{2}\int_0^\pi(1 - \cos 2x)\,dx \qquad u = 2x; \quad du = 2\,dx$$

145

$$V = \frac{\pi}{2} \int_0^\pi dx - \frac{\pi}{2} \frac{1}{2} \int_0^\pi \cos 2x (2\,dx) = \frac{\pi}{2} x \Big|_0^\pi - \frac{\pi}{4} \sin 2x \Big|_0^\pi$$

$$= \frac{\pi^2}{2} - 0 = \frac{\pi^2}{2}$$

41. Since
$$\sin\left(2t - \frac{\pi}{3}\right) = \sin 2t \cos \frac{\pi}{3} - \cos 2t \sin \frac{\pi}{3}$$

$$= \frac{1}{2} \sin 2t - \frac{\sqrt{3}}{2} \cos 2t,$$

we get, by Exercise 40 with $T = \pi$:

$$P = \frac{1}{\pi} \int_0^\pi (3 \sin 2t)(5)\left(\frac{1}{2} \sin 2t - \frac{\sqrt{3}}{2} \cos 2t\right) dt$$

$$= \frac{15}{2\pi} \int_0^\pi \sin^2 2t - \frac{15\sqrt{3}}{2\pi} \frac{1}{2} \int_0^\pi \sin 2t \cos 2t (2\,dt)$$

$$P = \frac{15}{2\pi} \frac{1}{2} \int_0^\pi (1 - \cos 4t) dt - \frac{15\sqrt{3}}{2\pi} \frac{1}{2} \int_0^\pi \sin 2t \cos 2t (2\,dt)$$

$$\qquad\qquad u = 4t; \ du = 4\,dt \qquad\qquad\qquad u = \sin 2t; \ du = 2 \cos 2t\,dt$$

$$P = \frac{15}{4\pi}\left(t - \frac{1}{4} \sin 4t\right)\Big|_0^\pi - \frac{15\sqrt{3}}{4\pi}\left(\frac{1}{2} \sin^2 2t\right)\Big|_0^\pi$$

$$= \frac{15}{4\pi}(\pi - 0) - 0 = \frac{15}{4} \ W$$

Section 7.5

1. $\int \dfrac{dx}{\sqrt{1 - x^2}} = \text{Arcsin } x + C$ by (7.19) with $a = 1$.

5. $\int \dfrac{x\,dx}{\sqrt{1 - x^2}} = \int (1 - x^2)^{-1/2}\,x\,dx \qquad\qquad u = 1 - x^2; \ du = -2x\,dx$

$$\qquad\qquad\qquad -\frac{1}{2} \int (1 - x^2)^{-1/2}(-2x\,dx)$$

$$= -\frac{1}{2} \int u^{-1/2}\,du = -\frac{1}{2} \frac{u^{1/2}}{1/2} + C = -\sqrt{u} + C$$

$$= -\sqrt{1 - x^2} + C$$

9. $\int \dfrac{t\,dt}{4 + t^4} = \int \dfrac{t\,dt}{4 + (t^2)^2} \qquad\qquad\qquad u = t^2; \ du = 2t\,dt$

$\frac{1}{2} \int \dfrac{2t\,dt}{4 + (t^2)^2} = \frac{1}{2} \int \dfrac{du}{4 + u^2} = \frac{1}{2} \frac{1}{2} \text{ Arctan } \frac{u}{2}$ by (7.20) with $a = 2$.

We now get:

$\frac{1}{4} \text{ Arctan } \frac{1}{2} t^2 + C$

13. $\displaystyle\int \frac{dx}{\sqrt{5 - 3x^2}} = \int \frac{dx}{\sqrt{5 - (\sqrt{3}x)^2}}$ $u = \sqrt{3}x; \quad du = \sqrt{3}\,dx$

$\displaystyle\frac{1}{\sqrt{3}} \int \frac{\sqrt{3}\,dx}{\sqrt{5 - (\sqrt{3}x)^2}} = \frac{1}{\sqrt{3}} \int \frac{du}{\sqrt{5 - u^2}} = \frac{1}{\sqrt{3}} \,\text{Arcsin}\,\frac{u}{\sqrt{5}} + C$

by (7.19) with $a = \sqrt{5}$.   We get:

$\displaystyle\frac{1}{\sqrt{3}} \,\text{Arcsin}\,\frac{\sqrt{3}x}{\sqrt{5}} + C = \frac{\sqrt{3}}{3} \,\text{Arcsin}\,\frac{\sqrt{15}x}{5} + C$

17. $\displaystyle\int_{3}^{3\sqrt{3}} \frac{3\,dx}{9 + x^2} = 3 \cdot \frac{1}{3} \,\text{Arctan}\,\frac{x}{3}\Big|_{3}^{3\sqrt{3}}$

$\displaystyle = \text{Arctan}\,\frac{3\sqrt{3}}{3} - \text{Arctan}\,\frac{3}{3}$

$\displaystyle = \text{Arctan}\,\sqrt{3} - \text{Arctan}\,1 = \frac{\pi}{3} - \frac{\pi}{4} = \frac{\pi}{12}$

21.   Completing the square, we have

$\quad x^2 - 6x + 10 = x^2 - 6x + \underline{9} - \underline{9} + 10 = (x^2 - 6x + 9) + 1$

$\qquad\qquad\qquad\quad = (x - 3)^2 + 1$

$\displaystyle\int \frac{dx}{x^2 - 6x + 10} = \int \frac{dx}{(x - 3)^2 + 1}$ $u = x - 3; \quad du = dx$

$\displaystyle\int \frac{du}{u^2 + 1} = \text{Arctan}\,u + C = \text{Arctan}(x - 3) + C$ by (7.20) with $a = 1$.

25.   Completing the square:

$\quad x^2 + 3x + 3 = x^2 + 3x + \frac{9}{4} - \frac{9}{4} + 3 = \left(x^2 + 3x + \frac{9}{4}\right) + \frac{3}{4}$

$\qquad\qquad\qquad\quad = \left(x + \frac{3}{2}\right)^2 + \frac{3}{4}$

$\displaystyle\int \frac{dx}{x^2 + 3x + 3} = \int \frac{dx}{\left(x + \frac{3}{2}\right)^2 + \frac{3}{4}}$ $u = x + \frac{3}{2}; \quad du = dx$

$\displaystyle\qquad\qquad = \int \frac{du}{u^2 + \frac{3}{4}} = \int \frac{du}{\left(\frac{\sqrt{3}}{2}\right)^2 + u^2}$

$\displaystyle\qquad\qquad = \frac{2}{\sqrt{3}} \,\text{Arctan}\,\frac{2}{\sqrt{3}}u + C$ by (7.20) with $a = \frac{\sqrt{3}}{2}$

We obtain

$\displaystyle\frac{2}{\sqrt{3}} \,\text{Arctan}\,\frac{2}{\sqrt{3}}\left(x + \frac{3}{2}\right) + C = \frac{2}{\sqrt{3}} \,\text{Arctan}\,\frac{2x + 3}{\sqrt{3}} + C$

29. $\displaystyle\int_{0}^{\pi/2} \frac{\cos x}{1 + \sin^2 x}\,dx$ $u = \sin x; \quad du = \cos x\,dx$

$\displaystyle\int_{0}^{\pi/2} \frac{\cos x}{1 + \sin^2 x} = \text{Arctan}\,(\sin x)\Big|_{0}^{\pi/2}$

$$= \text{Arctan } 1 - \text{Arctan } 0 = \frac{\pi}{4}$$

33. $\displaystyle\int \frac{e^\theta \, d\theta}{\sqrt{1 - e^{2\theta}}} = \int \frac{e^\theta \, d\theta}{\sqrt{1 - (e^\theta)^2}}$ $\qquad\qquad u = e^\theta; \ du = e^\theta \, d\theta$

$\displaystyle\int \frac{du}{\sqrt{1 - u^2}} = \text{Arcsin } u + C = \text{Arcsin } e^\theta + C$ by (7.19) with $a = 1$.

37. Volume of shell: $\quad 2\pi \cdot (\text{radius}) \cdot (\text{height}) \cdot (\text{thickness})$
$$= 2\pi x y \, dx$$

$\displaystyle V = 2\pi \int_0^\infty x \, \frac{1}{4 + x^4} dx = 2\pi \lim_{b\to\infty} \int_0^b \frac{x \, dx}{4 + (x^2)^2}$

$\displaystyle = 2\pi \lim_{b\to\infty} \frac{1}{2} \int_0^b \frac{2x \, dx}{4 + (x^2)^2} = \pi \lim_{b\to\infty} \frac{1}{2} \text{ Arctan } \frac{x^2}{2} \Big|_0^b$

$\displaystyle = \frac{\pi}{2} \lim_{b\to\infty} \left(\text{Arctan } \frac{b^2}{2} - 0 \right)$

Since $\tan \theta$ is undefined for $\theta = \frac{\pi}{2}$, we obtain:

$V = \frac{\pi}{2}\left(\frac{\pi}{2}\right) = \frac{\pi^2}{4}$

Section 7.6

1. $\displaystyle\int \frac{\sqrt{4 - x^2}}{x^2} dx.$ Let $x = 2 \sin \theta$, $dx = 2 \cos \theta \, d\theta$. Note that

$\sqrt{4 - x^2} = \sqrt{4 - 4 \sin^2\theta} = \sqrt{4(1 - \sin^2\theta} = 2\sqrt{\cos^2\theta} = 2 \cos \theta$

$\displaystyle\int \frac{\sqrt{4 - x^2} \, dx}{x^2} = \int \frac{(2 \cos \theta)(2 \cos \theta \, d\theta)}{4 \sin^2\theta} = \int \frac{\cos^2\theta}{\sin^2\theta} d\theta$

$\displaystyle\qquad\qquad = \int \cot^2\theta \, d\theta = \int (\csc^2\theta - 1) d\theta = -\cot \theta - \theta + C$

From the substituted expression $x = 2 \sin \theta$, we get
$\sin \theta = \frac{x}{2} \quad$ or $\quad \theta = \text{Arcsin } \frac{x}{2}$
To change $\cot \theta$, we use a diagram. Note that the opposite side has
length x and the hypotenuse length 2. The remaining side is of
length $\sqrt{4 - x^2}$ by the Pythagorean theorem.

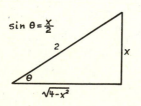
sin $\theta = \frac{x}{2}$

Since $\cot \theta = \dfrac{\sqrt{4 - x^2}}{x}$, we get

$\displaystyle\int \frac{\sqrt{4 - x^2}}{x^2} dx = -\cot \theta - \theta + C$

$\displaystyle\qquad = - \frac{\sqrt{4 - x^2}}{x} - \text{Arcsin } \frac{x}{2} + C$

5. $\int \dfrac{dx}{(x^2 + 25)^{3/2}}$.   Let $x = 5 \tan \theta$; $dx = 5 \sec^2 \theta d\theta$.   Then

$$(x^2 + 25)^{3/2} = (25 \tan^2 \theta + 25)^{3/2} = (25)^{3/2}(\tan^2 \theta + 1)^{3/2}$$

$$= 125(\sec^2 \theta)^{3/2} = 125 \sec^3 \theta$$

and we get:

$$\int \frac{5 \sec^2 \theta d\theta}{125 \sec^3 \theta} = \frac{1}{25} \int \frac{d\theta}{\sec \theta} = \frac{1}{25} \int \cos \theta d\theta = \frac{1}{25} \sin \theta + C$$

The diagram is constructed from the substituted expression

$\tan \theta = \dfrac{x}{5}$

Thus

$\dfrac{1}{25} \sin \theta + C$

$= \dfrac{1}{25} \dfrac{x}{\sqrt{x^2 + 25}} + C$

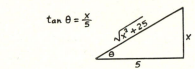

9. $\int \dfrac{dx}{x^2 \sqrt{x^2 + 16}}$.   Let $x = 4 \tan \theta$; $dx = 4 \sec^2 \theta d\theta$.   Then

$$\sqrt{x^2 + 16} = \sqrt{16 \tan^2 \theta + 16} = 4\sqrt{\tan^2 \theta + 1} = 4\sqrt{\sec^2 \theta} = 4 \sec \theta$$

$$\int \frac{dx}{x^2 \sqrt{x^2 + 16}} = \int \frac{4 \sec^2 \theta \; d\theta}{(16 \tan^2 \theta)(4 \sec \theta)} = \frac{1}{16} \int \frac{\sec \theta d\theta}{\tan^2 \theta} = \frac{1}{16} \int \frac{d\theta}{\cos \theta \tan^2 \theta}$$

$$= \frac{1}{16} \int \frac{\cos^2 \theta d\theta}{\cos \theta \sin^2 \theta} = \frac{1}{16} \int (\sin \theta)^{-2} \cos \theta d\theta$$

$$u = \sin \theta; \; du = \cos \theta d\theta$$

$$= \frac{1}{16} \int u^{-2} du + C = \frac{1}{16} \frac{u^{-1}}{-1} + C$$

$$= -\frac{1}{16} \frac{1}{\sin \theta} + C = -\frac{1}{16} \csc \theta + C$$

From $\tan \theta = \dfrac{x}{4}$, we construct the diagram shown at the right.

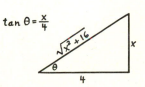

We now have

$$-\frac{1}{16} \csc \theta + C = -\frac{1}{16} \frac{\sqrt{x^2 + 16}}{x} + C$$

13. $\int \dfrac{x^3 \; dx}{\sqrt{x^2 - 3}}$.   Let $x = \sqrt{3} \sec \theta$; $dx = \sqrt{3} \sec \theta \tan \theta d\theta$.   Then

$$\sqrt{x^2 - 3} = \sqrt{3 \sec^2 \theta - 3} = \sqrt{3}\sqrt{\sec^2 \theta - 1} = \sqrt{3} \tan \theta$$

$$\int \frac{x^3 \, dx}{\sqrt{x^2 - 3}} = \int \frac{(\sqrt{3})^3 \sec^3\theta (\sqrt{3} \sec\theta \tan\theta \, d\theta)}{\sqrt{3} \tan\theta}$$

$$= 3\sqrt{3} \int \sec^4\theta \, d\theta = 3\sqrt{3} \int \sec^2\theta \, \sec^2\theta \, d\theta$$

$$= 3\sqrt{3} \int (1 + \tan^2\theta) \sec^2\theta \, d\theta \qquad\qquad u = \tan\theta; \quad du = \sec^2\theta \, d\theta$$

$$= 3\sqrt{3} \int (1 + u^2) \, du = 3\sqrt{3}\left(u + \tfrac{1}{3}u^3\right) + C$$

$$= 3\sqrt{3}\left(\tan\theta + \tfrac{1}{3}\tan^3\theta\right) + C$$

To construct the diagram, we use

$$\sec\theta = \frac{x}{\sqrt{3}}$$

$\sec\theta = \dfrac{x}{\sqrt{3}}$

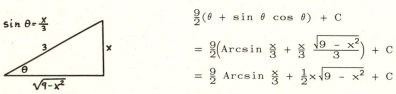

$$3\sqrt{3}\left(\tan\theta + \tfrac{1}{3}\tan^3\theta\right) = 3\sqrt{3}\left[\frac{\sqrt{x^2 - 3}}{\sqrt{3}} + \frac{1}{3}\left(\frac{\sqrt{x^2 - 3}}{\sqrt{3}}\right)^3\right] + C$$

$$= 3\sqrt{x^2 - 3} + \sqrt{3} \, \frac{(x^2 - 3)^{3/2}}{3\sqrt{3}} + C$$

$$= 3\sqrt{x^2 - 3} + \tfrac{1}{3}(x^2 - 3)^{3/2} + C$$

$$= \sqrt{x^2 - 3}\left[3 + \tfrac{1}{3}(x^2 - 3)\right] + C$$

$$= \tfrac{1}{3}(x^2 + 6)\sqrt{x^2 - 3} + C$$

19. $\int \sqrt{9 - x^2} \, dx$. Let $x = 3 \sin\theta$; $dx = 3 \cos\theta \, d\theta$. Then

$$\sqrt{9 - x^2} = \sqrt{9 - 9\sin^2\theta} = 3\sqrt{1 - \sin^2\theta} = 3 \cos\theta$$

$$\int \sqrt{9 - x^2} \, dx = \int (3 \cos\theta)(3 \cos\theta \, d\theta) = 9 \int \cos^2\theta \, d\theta$$

$$= \tfrac{9}{2} \int (1 + \cos 2\theta) \, d\theta = \tfrac{9}{2}\left(\theta + \tfrac{1}{2} \sin 2\theta\right) + C$$

Since $\sin 2\theta = 2 \sin\theta \cos\theta$, we get:

$\sin\theta = \dfrac{x}{3}$

$$\tfrac{9}{2}(\theta + \sin\theta \cos\theta) + C$$

$$= \tfrac{9}{2}\left(\text{Arcsin } \tfrac{x}{3} + \tfrac{x}{3}\frac{\sqrt{9 - x^2}}{3}\right) + C$$

$$= \tfrac{9}{2} \text{ Arcsin } \tfrac{x}{3} + \tfrac{1}{2}x\sqrt{9 - x^2} + C$$

For the definite integral, we now get:

$$\int_0^3 \sqrt{9 - x^2}\,dx = \frac{9}{2}\ \text{Arcsin}\ \frac{x}{3} + \frac{1}{2}x\sqrt{9 - x^2}\ \Big|_0^3$$

$$= \frac{9}{2}\ \text{Arcsin}\ 1 = \frac{9}{2}\ \frac{\pi}{2} = \frac{9\pi}{4}$$

21.

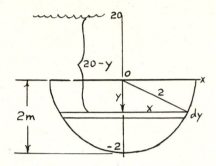

Length of strip: $2x = 2\sqrt{4 - y^2}$
Area of strip: $2\sqrt{4 - y^2}\ dy$
Pressure on strip: $(20 - y)w$
Force against strip: $(20 - y)w(2\sqrt{4 - y^2}\,dy)$
Summing from $y = -2$ to $y = 0$, we get:

$$F = \int_{-2}^0 2w(20 - y)\sqrt{4 - y^2}\,dy$$

$$= 2w \int_{-2}^0 20\sqrt{4 - y^2}\ dy - 2w \int_{-2}^0 y\sqrt{4 - y^2}\,dy$$

To evaluate $\int \sqrt{4 - y^2}\,dy$, we let $y = 2\ \sin\ \theta$, so that $dy = 2\ \cos\ \theta\,d\theta$.

$$\sqrt{4 - y^2} = \sqrt{4 - 4\ \sin^2\theta} = 2\sqrt{1 - \sin^2\theta} = 2\ \cos\ \theta$$

$$\int \sqrt{4 - y^2}\,dy = \int (2\ \cos\ \theta)(2\ \cos\ \theta\,d\theta) = 4 \int \cos^2\theta\,d\theta$$

$$= 4 \int \frac{1}{2}(1 + \cos\ 2\theta)\,d\theta = 2\Big(\theta + \frac{1}{2}\ \sin\ 2\theta\Big)$$

$$= 2(\theta + \sin\ \theta\ \cos\ \theta)$$

$$= 2\ \text{Arcsin}\ \frac{y}{2} + 2\Big(\frac{y}{2}\Big)\frac{\sqrt{4 - y^2}}{2}$$

$$= 2\ \text{Arcsin}\ \frac{y}{2} + \frac{1}{2}y\sqrt{4 - y^2}$$

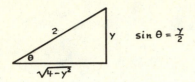

To evaluate $\int y\sqrt{4 - y^2}\,dy$, let $u = 4 - y^2$; $du = -2y\,dy$. Then

$$-\frac{1}{2}\int (4 - y^2)^{1/2}(-2y)\,dy = -\frac{1}{2}\frac{2}{3}(4 - y^2)^{3/2} = -\frac{1}{3}(4 - y^2)^{3/2}$$

We now have

$$F = 40w\left[2\,\text{Arcsin}\,\frac{y}{2} + \frac{1}{2}y\sqrt{4 - y^2}\right] - 2w\left[-\frac{1}{3}(4 - y^2)^{3/2}\right]_{-2}^{0}$$

$$= -40w\left[2 \cdot \left(-\frac{\pi}{2}\right)\right] - 2w\left(-\frac{1}{3} \cdot 4^{3/2}\right) \qquad \left(\text{since Arcsin}\,(-1) = -\frac{\pi}{2}\right)$$

$$= 40w\pi + \left(\frac{16}{3}\right)w = \left(40\pi + \frac{16}{3}\right)w \text{ N}$$

25.

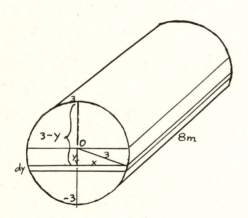

Volume of element: $2x(8dy) = 2\sqrt{9 - y^2}(8dy) = 16\sqrt{9 - y^2}\,dy$

Weight of element: $16w\sqrt{9 - y^2}\,dy$

Work done in moving a typical element to the top of the tank:

$(3 - y)(16w\sqrt{9 - y^2}\,dy)$

Summing from $y = -3$ to $y = 0$, we get:

$$W = 16w \int_{-3}^{0} (3 - y) \sqrt{9 - y^2} \, dy$$

$$= (16w) 3 \int_{-3}^{0} \sqrt{9 - y^2} \, dy - 16w \int_{-3}^{0} y \sqrt{9 - y^2} \, dy$$

The first integral has already been evaluated in Exercise 19. In the second integral, we let $u = 9 - y^2$; $du = -2y \, dy$. Thus

$$W = 48w \left(\frac{9}{2} \text{Arcsin } \frac{y}{3} + \frac{1}{2} y \sqrt{9 - y^2} \right) \Big|_{-3}^{0} - \frac{16w}{-2} \int_{-3}^{0} (9 - y^2)^{1/2} (-2y) \, dy$$

$$= 48w \left[0 - \frac{9}{2} \text{Arcsin } (-1) \right] + \left[8w \left(\frac{2}{3} \right) (9 - y^2)^{3/2} \right]_{-3}^{0}$$

$$= 48w \left[-\frac{9}{2} \left(-\frac{\pi}{2} \right) \right] + 8w \left(\frac{2}{3} \right) (9)^{3/2}$$

$$= 108\pi w + 144w = (108\pi + 144) w \text{ J}$$

Section 7.7

1. $\int x e^x \, dx$ $u = x$ $dv = e^x \, dx$

                                 $du = dx$ $v = e^x$

    $\int x e^x \, dx = x e^x - \int e^x \, dx = x e^x - e^x + C$

5. $\int x \sec^2 x \, dx$ $u = x$ $dv = \sec^2 x \, dx$

                                      $du = dx$ $v = \tan x$

    $\int x \sec^2 x \, dx = x \tan x - \int \tan x \, dx$

    $= x \tan x + \ln |\cos x| + C$

9. $\int \text{Arcsin } x \, dx$ $u = \text{Arcsin } x$ $dv = dx$

                                            $du = \dfrac{dx}{\sqrt{1 - x^2}}$ $v = x$

    $\int \text{Arcsin } x \, dx = x \text{ Arcsin } x - \int \dfrac{x}{\sqrt{1 - x^2}} \, dx$

    $= x \text{ Arcsin } x - \int (1 - x^2)^{-1/2} x \, dx$

    $= x \text{ Arcsin } x + \frac{1}{2} \int (1 - x^2)^{-1/2} (-2x) \, dx$ $u = 1 - x^2$; $du = -2x \, dx$

    $= x \text{ Arcsin } x + \frac{1}{2} \dfrac{(1 - x^2)^{1/2}}{1/2} + C$

    $= x \text{ Arcsin } x + \sqrt{1 - x^2} + C$

153

13. $\int x^2 e^{-x} dx$ $\qquad$ $u = x^2 \qquad dv = e^{-x} dx$

$\qquad\qquad\qquad\qquad\qquad\qquad\qquad du = 2x dx \qquad v = -e^{-x}$

$\int x^2 e^{-x} dx = -x^2 e^{-x} + 2 \int x e^{-x} dx$

$\qquad\qquad\qquad\qquad\qquad\qquad\qquad u = x \qquad dv = e^{-x} dx$

$\qquad\qquad\qquad\qquad\qquad\qquad\qquad du = dx \qquad v = -e^{-x}$

$\int x^2 e^{-x} dx = -x^2 e^{-x} + 2\left[-x e^{-x} + \int e^{-x} dx \right]$

$\qquad\qquad\qquad = -x^2 e^{-x} - 2x e^{-x} - 2e^{-x} + C$

$\qquad\qquad\qquad = -e^{-x}(x^2 + 2x + 2) + C$

17. $\int x \sin x^2 dx$. Here integration by parts is not needed: let $u = x^2$; $du = 2x dx$. Then

$\int x \sin x^2 dx = \frac{1}{2} \int (\sin x^2)(2x) dx = \frac{1}{2} \int \sin u \, du$

$\qquad\qquad\qquad = -\frac{1}{2} \cos u + C = -\frac{1}{2} \cos x^2 + C$

21. $\int e^{-x} \cos \pi x dx$ (see also Example 5)

$\qquad\qquad\qquad\qquad\qquad\qquad u = e^{-x} \qquad dv = \cos \pi x dx$

$\qquad\qquad\qquad\qquad\qquad\qquad du = -e^{-x} dx \qquad v = \frac{1}{\pi} \sin \pi x$

$\int e^{-x} \cos \pi x dx = \frac{1}{\pi} e^{-x} \sin \pi x + \frac{1}{\pi}\left[\int e^{-x} \sin \pi x dx \right]$

$\qquad\qquad\qquad\qquad\qquad\qquad u = e^{-x} \qquad dv = \sin \pi x dx$

$\qquad\qquad\qquad\qquad\qquad\qquad du = -e^{-x} dx \qquad v = -\frac{1}{\pi} \cos \pi x$

$\int e^{-x} \cos \pi x dx = \frac{1}{\pi} e^{-x} \sin \pi x + \frac{1}{\pi}\left[-\frac{1}{\pi} e^{-x} \cos \pi x - \frac{1}{\pi} \int e^{-x} \cos \pi x dx \right]$

$\qquad = \frac{1}{\pi} e^{-x} \sin \pi x - \frac{1}{\pi^2} e^{-x} \cos \pi x - \frac{1}{\pi^2} \int e^{-x} \cos \pi x dx$

Now solve for the integral:

$\int e^{-x} \cos \pi x dx + \frac{1}{\pi^2} \int e^{-x} \cos \pi x dx = \frac{1}{\pi} e^{-x} \sin \pi x - \frac{1}{\pi^2} e^{-x} \cos \pi x$

$\qquad\qquad \left(1 + \frac{1}{\pi^2} \right) \int e^{-x} \cos \pi x dx = \frac{1}{\pi^2} e^{-x}(\pi \sin \pi x - \cos \pi x)$

$\qquad\qquad \left(\frac{\pi^2 + 1}{\pi^2} \right) \int e^{-x} \cos \pi x dx = \frac{1}{\pi^2} e^{-x}(\pi \sin \pi x - \cos \pi x)$

We conclude that

$$\int e^{-x} \cos \pi x \, dx = \frac{1}{\pi^2 + 1} [e^{-x}(\pi \sin \pi x - \cos \pi x)] + C$$

25. $A = \int_0^\pi x \sin x \, dx$ $u = x$ $dv = \sin x \, dx$

                                           $du = dx$ $v = -\cos x$

$A = -x \cos x \big|_0^\pi + \int_0^\pi \cos x \, dx$

$ = -x \cos x \big|_0^\pi + \sin x \big|_0^\pi$

$ = -\pi \cos \pi + 0 = -\pi(-1) = \pi$

29. Recall that $v = \dfrac{q}{C} = \dfrac{1}{C} \int i \, dt$

$ = \dfrac{1}{10 \times 10^{-6}} \int e^{-t} \sin 4t \, dt = 10^5 \int e^{-t} \sin 4t \, dt$ (1)

$\int e^{-t} \sin 4t \, dt:$ $u = e^{-t}$ $dv = \sin 4t \, dt$

                                   $du = -e^{-t} dt$ $v = -\frac{1}{4} \cos 4t$

$\int e^{-t} \sin 4t \, dt = -\frac{1}{4} e^{-t} \cos 4t - \frac{1}{4}\left[\int e^{-t} \cos 4t \, dt \right]$

                                   $u = e^{-t}$ $dv = \cos 4t \, dt$

                                   $du = -e^{-t} dt$ $v = \frac{1}{4} \sin 4t$

$\int e^{-t} \sin 4t \, dt$

$ = -\frac{1}{4} e^{-t} \cos 4t - \frac{1}{4}\left[\frac{1}{4} e^{-t} \sin 4t + \frac{1}{4} \int e^{-t} \sin 4t \, dt \right]$

$ = -\frac{1}{4} e^{-t} \cos 4t - \frac{1}{16} e^{-t} \sin 4t - \frac{1}{16} \int e^{-t} \sin 4t \, dt$

Solving for the integral, we get:

$\frac{17}{16} \int e^{-t} \sin 4t \, dt = -\frac{1}{16} e^{-t}(4 \cos 4t + \sin 4t)$

$\phantom{\frac{17}{16}} \int e^{-t} \sin 4t \, dt = -\frac{1}{17} e^{-t}(4 \cos 4t + \sin 4t)$

By (1),

$v = 10^5 \left(-\frac{1}{17} e^{-t} \right)(4 \cos 4t + \sin 4t) + C$

If $t = 0$, $v = 0$. Thus

$0 = 10^5 \left(-\frac{1}{17} \right)(4) + C$ or $C = \dfrac{4(10^5)}{17}$

and

$$v = 10^5\left(-\frac{1}{17}e^{-t}\right)(4 \cos 4t + \sin 4t) + \left.\frac{4(10)^5}{17}\right|_{t=0.05}$$

$$= 482 \text{ V}$$

Section 7.8

1. $\displaystyle\int \frac{dx}{x^2 - 4}$

$$\frac{1}{x^2 - 4} = \frac{1}{(x - 2)(x + 2)} = \frac{A}{x - 2} + \frac{B}{x + 2} = \frac{A(x + 2) + B(x - 2)}{(x - 2)(x + 2)}$$

$A(x + 2) + B(x - 2) = 1$

<u>x = -2</u>: $0 + B(-4) = 1$ or $B = -\frac{1}{4}$

<u>x = 2</u>: $A(4) + 0 = 1$ or $A = \frac{1}{4}$

$$\int \frac{dx}{x^2 - 4} = \int\left(\frac{1}{4} \frac{1}{x - 2} - \frac{1}{4} \frac{1}{x + 2}\right)dx$$

$$= \frac{1}{4} \ln|x - 2| - \frac{1}{4} \ln|x + 2| + C = \frac{1}{4} \ln\left|\frac{x - 2}{x + 2}\right| + C$$

5. $\displaystyle\int \frac{5x - 4}{(x - 2)(x + 1)}dx$

$$\frac{5x - 4}{(x - 2)(x + 1)} = \frac{A}{x - 2} + \frac{B}{x + 1} = \frac{A(x + 1) + B(x - 2)}{(x - 2)(x + 1)}$$

$A(x + 1) + B(x - 2) = 5x - 4$

<u>x = -1</u>: $0 + B(-3) = -9$ or $B = 3$

<u>x = 2</u>: $A(3) + 0 = 6$ or $A = 2$

$$\int \frac{5x - 4}{(x - 2)(x + 1)}dx = \int\left(\frac{2}{x - 2} + \frac{3}{x + 1}\right)dx$$

$$= 2 \ln|x - 2| + 3 \ln|x + 1| + C$$

$$= \ln|x - 2|^2 + \ln|x + 1|^3 + C$$

$$= \ln|(x - 2)^2(x + 1)^3| + C$$

9. $\displaystyle\int \frac{x^2 + 10x - 20}{(x - 4)(x - 1)(x + 2)}dx$

$$\frac{x^2 + 10x - 20}{(x - 4)(x - 1)(x + 2)} = \frac{A}{x - 4} + \frac{B}{x - 1} + \frac{C}{x + 2}$$

$$= \frac{A(x - 1)(x + 2) + B(x - 4)(x + 2) + C(x - 4)(x - 1)}{(x - 4)(x - 1)(x + 2)}$$

$A(x - 1)(x + 2) + B(x - 4)(x + 2) + C(x - 4)(x - 1) = x^2 + 10x - 20$

<u>x = 1</u>: $0 + B(-3)(3) + 0 = -9$ or $B = 1$

<u>x = -2</u>: $0 + 0 + C(-6)(-3) = -36$ or $C = -2$

<u>x = 4</u>: $A(3)(6) + 0 + 0 = 36$ or $A = 2$

$$\int \frac{x^2 + 10x - 20}{(x - 4)(x - 1)(x + 2)}\,dx = \int \left(\frac{2}{x - 4} + \frac{1}{x - 1} + \frac{-2}{x + 2}\right)dx$$

$$= 2\ln|x - 4| + \ln|x - 1| - 2\ln|x + 2| + C$$
$$= \ln|x - 4|^2 + \ln|x - 1| - \ln|x + 2|^2 + C$$
$$= \ln\left|\frac{(x - 4)^2(x - 1)}{(x + 2)^2}\right| + C$$

13. $\int \dfrac{-2x^2 + 9x - 7}{(x - 2)^2(x + 1)}\,dx$

$$\frac{-2x^2 + 9x - 7}{(x - 2)^2(x + 1)} = \frac{A}{x - 2} + \frac{B}{(x - 2)^2} + \frac{C}{x + 1}$$
$$= \frac{A(x - 2)(x + 1) + B(x + 1) + C(x - 2)^2}{(x - 2)^2(x + 1)}$$

$A(x - 2)(x + 1) + B(x + 1) + C(x - 2)^2 = -2x^2 + 9x - 7$

<u>x = 2</u>: $0 + B(3) + 0 = 3$ or $B = 1$

<u>x = -1</u>: $0 + 0 + C(9) = -18$ or $C = -2$

    Using the values obtained for B and C, we get

    $A(x - 2)(x + 1) + 1(x + 1) - 2(x - 2)^2 = -2x^2 + 9x - 7$

<u>x = 0</u>: $A(-2)(1) + (1) - 2(4) = -7$ or $A = 0$

$\int \left(\dfrac{1}{(x - 2)^2} + \dfrac{-2}{x + 1}\right)dx$

$= \int \left[(x - 2)^{-2} - \dfrac{2}{x + 1}\right]dx$ $u = x - 2;\ du = dx$

$= \dfrac{(x - 2)^{-1}}{-1} - 2\ln|x + 1| + C$

$= -\dfrac{1}{x - 2} - 2\ln|x + 1| + C$

17. $\int \dfrac{x^2 - 3x - 2}{(x + 2)(x^2 + 4)}\,dx$

$$\frac{x^2 - 3x - 2}{(x + 2)(x^2 + 4)} = \frac{A}{x + 2} + \frac{Bx + C}{x^2 + 4}$$
$$= \frac{A(x^2 + 4) + (Bx + C)(x + 2)}{(x + 2)(x^2 + 4)}$$

$A(x^2 + 4) + (Bx + C)(x + 2) = x^2 - 3x - 2$

<u>x = -2</u>: $A(8) + 0 = 8$ or $A = 1$

  $1(x^2 + 4) + (Bx + C)(x + 2) = x^2 - 3x - 2$ $A = 1$

Now let x = 0 and solve for C:

  $(4) + C(2) = -2$ or $C = -3$ $x = 0$

Since A = 1 and C = -3, we get

$$1(x^2 + 4) + (Bx - 3)(x + 2) = x^2 - 3x - 2$$

At this point we may let x be equal to any value (such as x = 1):

$$5 + (B - 3)(3) = -4$$

or B = 0

So the integral becomes

$$\int \left(\frac{1}{x + 2} + \frac{-3}{x^2 + 4} \right) dx$$

$$= \ln |x + 2| - 3 \cdot \frac{1}{2} \text{ Arctan } \frac{x}{2} + C$$

$$= \ln |x + 2| - \frac{3}{2} \text{ Arctan } \frac{x}{2} + C$$

21. $$\int \frac{x^5 \, dx}{(x^2 + 4)^2} = \int \frac{x^5 \, dx}{x^4 + 8x^2 + 16}$$

$$x^4 + 8x^2 + 16 \overline{\smash{\big)}\ \begin{array}{l} x \\ x^5 \\ \underline{x^5 + 8x^3 + 16x} \\ - 8x^3 - 16x \end{array}}$$

The integrand can therefore be written as

$$x + \frac{-8x^3 - 16x}{(x^2 + 4)^2}$$

The fractional part has the form

$$\frac{-8x^3 - 16x}{(x^2 + 4)^2} = \frac{Ax + B}{x^2 + 4} + \frac{Cx + D}{(x^2 + 4)^2}$$

$$= \frac{(Ax + B)(x^2 + 4) + (Cx + D)}{(x^2 + 4)^2}$$

Equating numerators:

$$(Ax + B)(x^2 + 4) + (Cx + D) = -8x^3 - 16x$$
$$Ax^3 + Bx^2 + 4Ax + 4B + Cx + D = -8x^3 - 16x$$
$$Ax^3 + Bx^2 + (4A + C)x + (4B + D) = -8x^3 - 16x$$

Comparing coefficients:

A = -8, B = 0, 4A + C = -16, 4B + D = 0
whence C = 16 and D = 0. Thus

$$\int \frac{x^5 \, dx}{(x^2 + 4)^2} = \int \left(x + \frac{-8x}{x^2 + 4} + \frac{16x}{(x^2 + 4)^2} \right) dx$$

$$= \int x \, dx - 8 \int \frac{x \, dx}{x^2 + 4} + 16 \int (x^2 + 4)^{-2} x \, dx$$

$$= \int x \, dx - \frac{8}{2} \int \frac{2x \, dx}{x^2 + 4} + \frac{16}{2} \int (x^2 + 4)^{-2} (2x \, dx)$$

$$u = x^2 + 4; \quad du = 2x \, dx$$

$$\int \frac{x^5 \, dx}{(x^2 + 4)^2} = \frac{1}{2}x^2 - 4\ln(x^2 + 4) + 8\frac{(x^2 + 4)^{-1}}{-1} + C$$

$$= \frac{1}{2}x^2 - 4\ln(x^2 + 4) - \frac{8}{x^2 + 4} + C$$

25. $\int \dfrac{dx}{x(x^2 + 2x + 2)}$

$$\frac{1}{x(x^2 + 2x + 2)} = \frac{A}{x} + \frac{Bx + C}{x^2 + 2x + 2}$$

$$= \frac{A(x^2 + 2x + 2) + (Bx + C)x}{x(x^2 + 2x + 2)}$$

$$A(x^2 + 2x + 2) + (Bx + C)x = 1$$

$$Ax^2 + 2Ax + 2A + Bx^2 + Cx = 1$$

$$(A + B)x^2 + (2A + C)x + 2A = 1$$

Comparing coefficients:

$\quad$ A + B = 0 $\quad$ coefficients of x^2

$\quad$ 2A + C = 0 $\quad$ coefficients of x

$\quad\quad$ 2A = 1 $\quad$ constants

solution set: $\quad A = \frac{1}{2}, \quad B = -\frac{1}{2}, \quad C = -1$

$$\int \frac{dx}{x(x^2 + 2x + 2)} = \int \left(\frac{1}{2}\frac{1}{x} + \frac{-\frac{1}{2}x - 1}{x^2 + 2x + 2} \right) dx$$

$$= \frac{1}{2} \int \frac{1}{x} dx - \frac{1}{2} \int \frac{x + 2}{x^2 + 2x + 2} dx$$

For the second integral, if $u = x^2 + 2x + 2$, then $du = (2x + 2)dx$. So the integral has to be split as follows:

$$\frac{1}{2} \int \frac{1}{x} dx - \frac{1}{2}\frac{1}{2} \int \frac{2(x + 2)}{x^2 + 2x + 2} dx$$

$$= \frac{1}{2} \int \frac{1}{x} dx - \frac{1}{4} \int \frac{2x + 4}{x^2 + 2x + 2} dx$$

$$= \frac{1}{2} \int \frac{1}{x} dx - \frac{1}{4} \int \frac{(2x + 2) + 2}{x^2 + 2x + 2} dx$$

$$= \frac{1}{2} \int \frac{1}{x} dx - \frac{1}{4} \int \frac{2x + 2}{x^2 + 2x + 2} dx - \frac{1}{4} \int \frac{2 \, dx}{x^2 + 2x + 2}$$

$$= \frac{1}{2} \int \frac{1}{x} dx - \frac{1}{4} \int \frac{2x + 2}{x^2 + 2x + 2} dx - \frac{1}{2} \int \frac{dx}{(x + 1)^2 + 1}$$

$$= \frac{1}{2} \ln|x| - \frac{1}{4} \ln|x^2 + 2x + 2| - \frac{1}{2} \text{Arctan}(x + 1) + C$$

Section 7.9

1. Formula 5; $a = 2$, $b = 1$

$$\int \frac{dx}{x(2 + x)} = \frac{1}{2} \ln \left| \frac{x}{2 + x} \right| + C$$

5. Formula 16; $a = \sqrt{5}$

$$\int \frac{dx}{5 - x^2} = \frac{1}{2\sqrt{5}} \ln \left| \frac{\sqrt{5} + x}{\sqrt{5} - x} \right| + C$$

9. Formula 61; $m = 2$, $n = 1$

$$\int \sin 2x \sin x\, dx = -\frac{\sin(2 + 1)x}{2(2 + 1)} + \frac{\sin(2 - 1)x}{2(2 - 1)} + C$$

$$= -\frac{1}{6} \sin 3x + \frac{1}{2} \sin x + C$$

13. Formula 42; $n = 2$, $a = 2$

$$\int x^2 e^{2x}\, dx = \frac{x^2 e^{2x}}{2} - \frac{2}{2} \int x e^{2x}\, dx = \frac{1}{2} x^2 e^{2x} - \int x e^{2x}\, dx$$

Now, by formula 41 with $a = 2$, we get:

$$\frac{1}{2} x^2 e^{2x} - \left[\frac{e^{2x}}{4}(2x - 1) \right] + C = \frac{1}{2} x^2 e^{2x} - \frac{1}{2} x e^{2x} + \frac{1}{4} e^{2x} + C$$

17. $$\int \frac{dx}{4x^2 - 9} = \int \frac{dx}{(2x)^2 - 9} = \frac{1}{2} \int \frac{du}{u^2 - 9} \qquad u = 2x;\ du = 2\,dx$$

By formula 17 with $a = 3$, we get:

$$\frac{1}{2} \frac{1}{6} \ln \left| \frac{u - 3}{u + 3} \right| + C = \frac{1}{12} \ln \left| \frac{2x - 3}{2x + 3} \right| + C$$

21. $$\int \frac{dx}{3x^2 - 5} = \int \frac{dx}{(\sqrt{3}x)^2 - 5} = \frac{1}{\sqrt{3}} \int \frac{\sqrt{3}\ dx}{(\sqrt{3}x)^2 - 5} \qquad u = \sqrt{3}x;\ du = \sqrt{3}\,dx$$

$$= \frac{1}{\sqrt{3}} \int \frac{du}{u^2 - 5}$$

By formula 17 with $a = \sqrt{5}$, we get:

$$\frac{1}{\sqrt{3}} \frac{1}{2\sqrt{5}} \ln \left| \frac{u - \sqrt{5}}{u + \sqrt{5}} \right| + C = \frac{1}{2\sqrt{15}} \ln \left| \frac{\sqrt{3}x - \sqrt{5}}{\sqrt{3}x + \sqrt{5}} \right| + C$$

25. Formula 63, $m = 3$, $n = 2$

$$\int \sin 3x \cos 2x\, dx = -\frac{\cos(3 + 2)x}{2(3 + 2)} - \frac{\cos(3 - 2)x}{2(3 - 2)}$$

$$= -\frac{1}{10} \cos 5x - \frac{1}{2} \cos x + C$$

1. $\displaystyle\int \frac{x\ dx}{x^2 + 1} = \frac{1}{2}\int \frac{2x\ dx}{x^2 + 1}$ $\qquad\qquad\qquad u = x^2 + 1;\ du = 2x\,dx$

$\displaystyle = \frac{1}{2}\ \ln|u| + C = \frac{1}{2}\ \ln(x^2 + 1) + C$

5. $\displaystyle\int \frac{t\ dt}{\sqrt{9 - t^2}} = \int (9 - t^2)^{-1/2}t\,dt$ $\qquad\qquad u = 9 - t^2;\ du = -2t\,dt$

$\displaystyle = -\frac{1}{2}\int (9 - t^2)^{-1/2}(-2t)\,dt$

$\displaystyle = -\frac{1}{2}\ \frac{(9 - t^2)^{1/2}}{1/2} + C = -\sqrt{9 - t^2} + C$

9. $\displaystyle\int \frac{e^x\ dx}{4 + e^{2x}} = \int \frac{e^x\ dx}{4 + (e^x)^2}$ $\qquad\qquad\qquad u = e^x;\ du = e^x dx$

$\displaystyle\int \frac{du}{4 + u^2} = \frac{1}{2}\ \text{Arctan}\ \frac{u}{2} + C = \frac{1}{2}\ \text{Arctan}\ \frac{1}{2}e^x + C$

13. $\displaystyle\int \frac{\ln x}{x}dx$ $\qquad\qquad\qquad\qquad\qquad u = \ln x;\ du = \frac{1}{x}dx$

$\displaystyle\int (\ln x)\frac{1}{x}dx = \int u\,du = \frac{1}{2}u^2 + C$

$\displaystyle = \frac{1}{2}\ \ln^2 x + C$

17. $\displaystyle\int \frac{dx}{x^2 + 4x + 4} = \int \frac{dx}{(x + 2)^2}$ $\qquad\qquad u = x + 2;\ du = dx$

$\displaystyle = \int (x + 2)^{-2}dx = \frac{(x + 2)^{-1}}{-1} + C$

$\displaystyle = -\frac{1}{x + 2} + C$

21. $\displaystyle\int \frac{dx}{x\sqrt{4 - x^2}}.$ Let $x = 2\sin\theta$; then $dx = 2\cos\theta\,d\theta$

$\sqrt{4 - x^2} = \sqrt{4 - 4\sin^2\theta} = 2\sqrt{1 - \sin^2\theta} = 2\sqrt{\cos^2\theta} = 2\cos\theta$

$\displaystyle\int \frac{2\cos\theta\,d\theta}{(2\sin\theta)(2\cos\theta)} = \frac{1}{2}\int \csc\theta\,d\theta = \frac{1}{2}\ \ln|\csc\theta - \cot\theta| + C$

$\displaystyle = \frac{1}{2}\ \ln\left|\frac{2}{x} - \frac{\sqrt{4 - x^2}}{x}\right| + C$

$\displaystyle = \frac{1}{2}\ \ln\left|\frac{2 - \sqrt{4 - x^2}}{x}\right| + C$

$\sin\theta = \frac{x}{2}$

161

25. $\int \sin^2 2x\,dx = \frac{1}{2}\int(1 - \cos 4x)dx = \frac{1}{2}\int dx - \frac{1}{2}\int\cos 4x\,dx$

$\qquad = \frac{1}{2}x - \frac{1}{2}\left(\frac{1}{4}\sin 4x\right) + C \qquad u = 4x; \ du = 4dx$

$\qquad = \frac{1}{2}x - \frac{1}{8}\sin 4x + C$

29. $\int \text{Arctan } 2x\,dx$ $\qquad\qquad\qquad u = \text{Arctan } 2x \qquad dv = dx$

$\qquad\qquad\qquad\qquad\qquad\qquad du = \dfrac{2\,dx}{1 + 4x^2} \qquad v = x$

$\quad x\ \text{Arctan } 2x - \int \dfrac{2x\,dx}{1 + 4x^2}$

$= x\ \text{Arctan } 2x - \frac{1}{4}\int \dfrac{8x\,dx}{1 + 4x^2} \qquad u = 1 + 4x^2; \ du = 8x\,dx$

$= x\ \text{Arctan } 2x - \frac{1}{4}\ln(1 + 4x^2) + C$

33. $(x^2 - 1) \div (x^2 + 3) = 1 - \dfrac{4}{x^2 + 3}$

$\int \dfrac{x^2 - 1}{x^2 + 3}dx = \int\left[1 - \dfrac{4}{x^2 + 3}\right]dx = x - \dfrac{4}{\sqrt{3}}\ \text{Arctan } \dfrac{x}{\sqrt{3}} + C$

37. $\int \cot^4 x\ \csc^4 x\,dx = \int \cot^4 x\ \csc^2 x(\csc^2 x\,dx)$

$\quad = \int \cot^4 x(1 + \cot^2 x)(\csc^2 x\,dx)$

$\quad = -\int \cot^4 x(1 + \cot^2 x)(-\csc^2 x\,dx) \qquad u = \cot x; \ du = -\csc^2 x\,dx$

$\quad = -\int u^4(1 + u^2)du = -\frac{1}{5}u^5 - \frac{1}{7}u^7 + C$

$\quad = -\frac{1}{5}\cot^5 x - \frac{1}{7}\cot^7 x + C$

41. $\int \tan^2 x\ \sec^2 x\,dx \qquad\qquad\qquad u = \tan x; \ du = \sec^2 x\,dx$

$\int u^2 du = \frac{1}{3}u^3 + C = \frac{1}{3}\tan^3 x + C$

45. $\int e^x\ \cos 4x\,dx: \qquad\qquad\qquad u = e^x \qquad dv = \cos 4x\,dx$

$\qquad\qquad\qquad\qquad\qquad\qquad du = e^x dx \qquad v = \frac{1}{4}\sin 4x$

$\int e^x \cos 4x\,dx = \frac{1}{4}e^x\sin 4x - \frac{1}{4}\int e^x\sin 4x\,dx$

$\qquad\qquad\qquad\qquad\qquad u = e^x \qquad dv = \sin 4x\,dx$

$\qquad\qquad\qquad\qquad\qquad du = e^x dx \qquad v = -\frac{1}{4}\cos 4x$

$$\int e^x \cos\ 4x\,dx = \tfrac{1}{4}e^x \sin\ 4x\ -\ \tfrac{1}{4}\left[-\tfrac{1}{4}e^x \cos\ 4x\ +\ \tfrac{1}{4}\int e^x \cos\ 4x\,dx\right]$$

$$= \tfrac{1}{4}e^x \sin\ 4x\ +\ \tfrac{1}{16}e^x \cos\ 4x\ -\ \tfrac{1}{16}\int e^x \cos\ 4x\,dx$$

Solving for the integral:

$$\left(1\ +\ \tfrac{1}{16}\right)\int e^x \cos\ 4x\,dx = \tfrac{1}{4}e^x \sin\ 4x\ +\ \tfrac{1}{16}e^x \cos\ 4x$$

$$\tfrac{17}{16}\int e^x \cos\ 4x\,dx = \tfrac{1}{16}e^x(4\ \sin\ 4x\ +\ \cos\ 4x)$$

We conclude that

$$\int e^x \cos\ 4x\,dx = \tfrac{1}{17}e^x(4\ \sin\ 4x\ +\ \cos\ 4x)\ +\ C$$

49. $\displaystyle\int \frac{x\ +\ 1}{x^2\ +\ 2x\ -\ 8}\,dx$ $\qquad\qquad u = x^2\ +\ 2x\ -\ 8;\ du = (2x\ +\ 2)\,dx$

$\displaystyle= \tfrac{1}{2}\int \frac{2(x\ +\ 1)}{x^2\ +\ 2x\ -\ 8}\,dx = \tfrac{1}{2}\ \ln|x^2\ +\ 2x\ -\ 8|\ +\ C$

CHAPTER 8   PARAMETRIC EQUATIONS, VECTORS, AND POLAR COORDINATES

Section 8.1

1. $x = 3t$, $y = t + 1$. From the second equation, $t = y - 1$. Substituting in the first equation, we get $x = 3(y - 1)$, and $x - 3y + 3 = 0$.

5.
$$x = \cos \theta$$
$$\underline{y = \sin^2\theta}$$
$$x^2 = \cos^2\theta$$
$$\underline{y = \sin^2\theta}$$
$$x^2 + y = \cos^2\theta + \sin^2\theta = 1 \text{ (adding)}$$
$$y = 1 - x^2$$

9. From $x = \ln t$, we get $e^x = t$. Substituting in $y = t + 2$, we have $y = e^x + 2$.

13. $x = e^t$; $v_x = e^t\big|_{t=0} = 1$

$y = e^{-t}$; $v_y = -e^{-t}\big|_{t=0} = -1$

Thus $\vec{v} = \vec{i} - \vec{j}$.

17. $x = 4 \cos t$; $v_x = -4 \sin t$, $a_x = -4 \cos t$

$y = 4 \sin t$; $v_y = 4 \cos t$, $a_y = -4 \sin t$

When $t = \dfrac{3\pi}{4}$, then

$v_x = -\dfrac{4}{\sqrt{2}}$, $a_x = \dfrac{4}{\sqrt{2}}$ and $v_y = -\dfrac{4}{\sqrt{2}}$, $a_y = -\dfrac{4}{\sqrt{2}}$

Thus

$\vec{v} = -\dfrac{4}{\sqrt{2}}\vec{i} - \dfrac{4}{\sqrt{2}}\vec{j}$ and $\vec{a} = \dfrac{4}{\sqrt{2}}\vec{i} - \dfrac{4}{\sqrt{2}}\vec{j}$

164

$$|\vec{v}| = |\vec{a}| = \sqrt{\frac{16}{2} + \frac{16}{2}} = \sqrt{16} = 4$$

The direction of $\vec{v}$ is 225° and the direction of $\vec{a}$ is 315°. The curve is a circle of radius 4 centered at the origin.

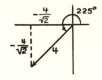

21. By Galileo's principle, the horizontal and vertical components of the velocity can be treated separately.

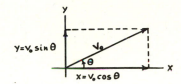

In the horizontal direction, we need only the relationship

distance = rate × time

to obtain

$x = (v_0 \cos \theta)t = v_0 t \cos \theta$.
In the vertical direction, we have $g = -10$ m/s^2 by the assumption that the upward direction is positive. So if v denotes the upward velocity, then $v = -10t + C$. If $t = 0$, $v = v_0 \sin \theta$, so that

$v_0 \sin \theta = 0 + C$. Thus

$v = -10t + v_0 \sin \theta$

Integrating, we get

$y = -5t^2 + (v_0 \sin \theta)t + k$

If $t = 0$, then $0 = k$, so that

$y = v_0 t \sin \theta - 5t^2$

(b) If $v_0 = 40$ m/s and $\theta = 30°$, then

$x = 40t \cos 30°, \quad y = 40t \sin 30° - 5t^2$

$x = 40t\left(\frac{\sqrt{3}}{2}\right), \quad y = 40t\left(\frac{1}{2}\right) - 5t^2$

$x = 20\sqrt{3}t, \quad y = 20t - 5t^2$

$v_x = 20\sqrt{3}, \quad v_y = 20 - 10t|_{t=1} = 10$

Magnitude: $\sqrt{(20\sqrt{3})^2 + (10)^2} = \sqrt{1200 + 100} = 10\sqrt{13}$ m/s

Direction: $\text{Arctan}\left[\dfrac{10}{20\sqrt{3}}\right] = 16.1°$

Also

$a_x = 0$, $a_y = -10$

Magnitude: $\sqrt{0 + 100} = 10$ m/s^2

Direction: $-90°$

(c) The range is the value of x when $y = 0$:

$$0 = v_0 t \sin\theta - 5t^2$$

$$t(v_0 \sin\theta - 5t) = 0$$

$$t = 0, \quad t = \tfrac{1}{5} v_0 \sin\theta$$

Substituting the nonzero value in the equation $x = v_0 t \cos\theta$, we get:

$$R = x = v_0\left(\tfrac{1}{5}v_0 \sin\theta\right)\cos\theta$$

$$R = \tfrac{1}{5} v_0^2 \sin\theta \cos\theta$$

Section 8.2

1. $y = \tfrac{2}{3}x^{3/2}$; $y' = \tfrac{2}{3}\,\tfrac{3}{2}x^{1/2} = x^{1/2}$; $(y')^2 = x$

$$s = \int_0^3 \sqrt{1 + x}\,dx = \tfrac{2}{3}(1 + x)^{3/2}\Big|_0^3 = \tfrac{2}{3}\left[4^{3/2} - 1\right] = \tfrac{14}{3}$$

5. $y = \tfrac{1}{6}x^3 + \tfrac{1}{2x} = \tfrac{1}{6}x^3 + \tfrac{1}{2}x^{-1}$

$y' = \tfrac{1}{2}x^2 - \tfrac{1}{2}x^{-2}$ $\qquad\qquad (y')^2 = \tfrac{1}{4}x^4 - \tfrac{1}{2} + \tfrac{1}{4}x^{-4}$

$1 + (y')^2 = 1 + \tfrac{1}{4}x^4 - \tfrac{1}{2} + \tfrac{1}{4}x^{-4} = \tfrac{1}{4}x^4 + \tfrac{1}{2} + \tfrac{1}{4}x^{-4} = \left(\tfrac{1}{2}x^2 + \tfrac{1}{2}x^{-2}\right)^2$

$$s = \int_1^3 \sqrt{\left(\tfrac{1}{2}x^2 + \tfrac{1}{2}x^{-2}\right)^2}\,dx = \int_1^3 \left(\tfrac{1}{2}x^2 + \tfrac{1}{2}x^{-2}\right)dx$$

$$= \tfrac{1}{2}\int_1^3 (x^2 + x^{-2})\,dx = \tfrac{1}{2}\left(\tfrac{1}{3}x^3 + \tfrac{x^{-1}}{-1}\right)\Big|_1^3 = \tfrac{1}{2}\left(\tfrac{1}{3}x^3 - \tfrac{1}{x}\right)\Big|_1^3$$

$$= \tfrac{1}{2}\left[\left(9 - \tfrac{1}{3}\right) - \left(\tfrac{1}{3} - 1\right)\right] = \tfrac{1}{2}\left(10 - \tfrac{2}{3}\right) = \tfrac{14}{3}$$

9. $x = \cos^3\theta$, $y = \sin^3\theta$

$\dfrac{dx}{d\theta} = 3\cos^2\theta(-\sin\theta)$, $\dfrac{dy}{d\theta} = 3\sin^2\theta\cos\theta$

$\left(\dfrac{dx}{d\theta}\right)^2 = 9\cos^4\theta\sin^2\theta$, $\left(\dfrac{dy}{d\theta}\right)^2 = 9\sin^4\theta\cos^2\theta$

$$s = \int_0^{\pi/2} \sqrt{9\cos^4\theta\sin^2\theta + 9\sin^4\theta\cos^2\theta}\,d\theta$$

$$= \int_0^{\pi/2} \sqrt{9\cos^2\theta\sin^2\theta(\cos^2\theta + \sin^2\theta)}\,d\theta$$

$$= \int_0^{\pi/2} 3 \cos\theta \sin\theta = 3 \int_0^{\pi/2} \sin\theta \cos\theta \, d\theta \qquad u = \sin\theta; \ du = \cos\theta$$

$$= \tfrac{3}{2} \sin^2\theta \Big|_0^{\pi/2} = \tfrac{3}{2}$$

13. $y = \left(a^{2/3} - x^{2/3}\right)^{3/2}$

$\quad y' = \tfrac{3}{2}\left(a^{2/3} - x^{2/3}\right)^{1/2}\left(-\tfrac{2}{3}x^{-1/3}\right)$

$\quad\quad = -x^{-1/3}\left(a^{2/3} - x^{2/3}\right)^{1/2}$

$\quad (y')^2 = \left(-x^{-1/3}\right)^2\left[\left(a^{2/3} - x^{2/3}\right)^{1/2}\right]^2$

$\quad\quad = x^{-2/3}\left(a^{2/3} - x^{2/3}\right)$

$\quad\quad = x^{-2/3}a^{2/3} - 1$

$\quad 1 + (y')^2 = x^{-2/3}a^{2/3}$

$\quad \sqrt{1 + (y')^2} = x^{-1/3}a^{1/3}$

Arc length $= 4 \int_0^a x^{-1/3}a^{1/3}\,dx = 4a^{1/3} \int_0^a x^{-1/3}\,dx$

$$= 4a^{1/3} \cdot \tfrac{3}{2}x^{2/3}\Big|_0^a = 6a^{1/3}a^{2/3} = 6a$$

Section 8.3

1.

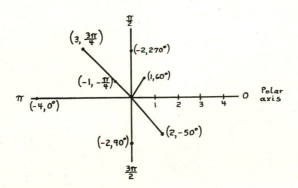

5. $x = r \cos\theta = -6 \cos\left(\tfrac{\pi}{3}\right) = -6\left(\tfrac{1}{2}\right) = -3$

$\quad y = r \sin\theta = -6 \sin\left(\tfrac{\pi}{3}\right) = -6\left(\tfrac{\sqrt{3}}{2}\right) = -3\sqrt{3}$

9. $r^2 = x^2 + y^2 = 1^2 + (-1)^2 = 2, \ r = \sqrt{2}$

$\quad \tan\theta = \tfrac{-1}{1} = -1.$ Since θ is in the fourth quadrant, $\theta = 315° = \tfrac{7\pi}{4}$

13. x = 2

   r cos θ = 2

   $$r = \frac{2}{\cos \theta}$$
   r = 2 sec θ

17. $x^2 - y^2 = 1$

   $(r \cos \theta)^2 - (r \sin \theta)^2 = 1$

   $r^2 \cos^2 \theta - r^2 \sin^2 \theta = 1$

   $r^2(\cos^2 \theta - \sin^2 \theta) = 1$

   $r^2(\cos 2\theta) = 1$ double-angle formula

   $$r^2 = \frac{1}{\cos 2\theta} = \sec 2\theta$$

21. r sin θ = 2 leads directly to y = 2.

25.   r = cos θ

   $r^2 = r \cos \theta$ multiplying by r

   $x^2 + y^2 = x$

   Alternatively,

   $r = \cos \theta = \frac{x}{r}$

   From $r = \frac{x}{r}$, we have $r^2 = x$ or $x^2 + y^2 = x$.

29. r = 1 - cos θ

   $r^2 = r - r \cos \theta$ multiplying by r

   $x^2 + y^2 = \pm\sqrt{x^2 + y^2} - x$

   $x^2 + y^2 + x = \pm\sqrt{x^2 + y^2}$

   $(x^2 + y^2 + x)^2 = x^2 + y^2$ squaring both sides

   Alternatively,

   r = 1 - cos θ

   $r = 1 - \frac{x}{r}$ $\cos \theta = \frac{x}{r}$

   $\pm\sqrt{x^2 + y^2} = 1 - \frac{x}{\pm\sqrt{x^2 + y^2}}$

   $x^2 + y^2 = \pm\sqrt{x^2 + y^2} - x$ multiplying by $\pm\sqrt{x^2 + y^2}$

33. $r = \frac{2}{1 - \cos \theta}$. Dividing both sides by r, we get

   $1 = \frac{2}{r(1 - \cos \theta)} = \frac{2}{r - r \cos \theta}$

   $1 = \frac{2}{\pm\sqrt{x^2 + y^2} - x}$

   $\pm\sqrt{x^2 + y^2} - x = 2$

168

$\pm \sqrt{x^2 + y^2} = x + 2$

$x^2 + y^2 = x^2 + 4x + 4$ squaring both sides

$y^2 = 4(x + 1)$

37. $r^2 = 3 \cos 2\theta = 3(\cos^2\theta - \sin^2\theta)$ double-angle formula

   $r^4 = 3(r^2\cos^2\theta - r^2\sin^2\theta)$ multiplying by r^2

   $(x^2 + y^2)^2 = 3(x^2 - y^2)$

39. $r = a \sin 3\theta = a \sin(2\theta + \theta)$

        $= a(\sin 2\theta \cos \theta + \cos 2\theta \sin \theta)$

   $r = a[(2 \sin \theta \cos \theta)\cos \theta + (\cos^2\theta - \sin^2\theta)\sin \theta]$

    $= a(2 \sin \theta \cos^2\theta + \cos^2\theta \sin \theta - \sin^3\theta)$

    $= a(3 \sin \theta \cos^2\theta - \sin^3\theta)$

   $r^4 = a[3(r \sin \theta)(r^2\cos^2\theta) - r^3\sin^3\theta]$ multiplying by r^3

   $(x^2 + y^2)^2 = a(3x^2y - y^3)$

Section 8.4

5. $r \cos \theta = 4$ is the line $x = 4$

9. Limacon. If $\theta = 90°$ or $270°$, $r = 9$

        If $\theta = 0°$, $r = 14$

        If $\theta = 180°$, $r = 4$

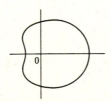

169

13. Limacon with a loop.

If $\theta = 0°$ or $180°$, $r = 1$

If $\theta = 90°$, $r = -1$

If $\theta = 270°$, $r = 3$

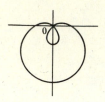

Also, $r = 0$ when

$1 - 2 \sin \theta = 0$

$\sin \theta = \frac{1}{2}$

$\theta = 30°, 150°$

17. Four-leaf rose. $r = 0$ when $\theta = 0°, 90°, 180°, 270°$

$r = 4$ when $\theta = 45°, 135°, 225°, 315°$

25. $r = \cos 3(\theta - \frac{\pi}{6})$ is the three-leaf rose $r = \cos 3\theta$ rotated $\frac{\pi}{6}$ in the counterclockwise direction.

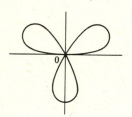

29. $r = \frac{2}{\theta}$

as $\theta \to 0$, $r \to \infty$

as $\theta \to \infty$, $r \to 0$

(In other words, as the curve circles the pole, r gets closer to 0; see curve in answer section.)

33. If $\theta = 0°$ or $180°$, $r = 3$

If $\theta = 90°$, $r = \frac{6}{5}$

If $\theta = 270°$, $r = -6$

Some other points are given in the following table:

θ:	$-10°$	$190°$	$250°$	$260°$	$280°$	$290°$
r:	4.1	4.1	-7.3	-6.3	-6.3	-7.3

170

The resulting curve is a hyperbola. (See the sketch in the answer section.)

Section 8.5

1. $A = \frac{1}{2} \int_0^{\pi/6} 2^2 \, d\theta = 2\theta \big|_0^{\pi/6} = \frac{\pi}{3}$

5. $A = \frac{1}{2} \int_0^2 \theta^2 \, d\theta = \frac{1}{2} \cdot \frac{1}{3} \theta^3 \big|_0^2 = \frac{4}{3}$

9. For a limacon without a loop, the limits of integration are 0 and 2π:

$$A = \frac{1}{2} \int_0^{2\pi} (2 + \cos\theta)^2 \, d\theta = \frac{1}{2} \int_0^{2\pi} (4 + 4\cos\theta + \cos^2\theta) \, d\theta$$

Since $\cos^2\theta = \frac{1}{2}(1 + \cos 2\theta)$, we get

$$A = \frac{1}{2} \int_0^{2\pi} \left[4 + 4\cos\theta + \frac{1}{2}(1 + \cos 2\theta) \right] d\theta$$

$$= \frac{1}{2} \left[4\theta + 4\sin\theta + \frac{1}{2}\left(\theta + \frac{1}{2}\sin 2\theta\right) \right]_0^{2\pi} \qquad u = 2\theta; \ du = 2 \, d\theta$$

$$= \frac{1}{2} \left[4(2\pi) + \frac{1}{2}(2\pi) \right] = \frac{9\pi}{2}$$

13. For a limacon without a loop, the limits of integration are 0 and 2π:

$$A = \frac{1}{2} \int_0^{2\pi} (2 - \sin\theta)^2 \, d\theta = \frac{1}{2} \int_0^{2\pi} (4 - 4\sin\theta + \sin^2\theta) \, d\theta$$

$$= \frac{1}{2} \int_0^{2\pi} \left[4 - 4\sin\theta + \frac{1}{2}(1 - \cos 2\theta) \right] d\theta$$

$$= \frac{1}{2} \left[4\theta + 4\cos\theta + \frac{1}{2}\left(\theta - \frac{1}{2}\sin 2\theta\right) \right]_0^{2\pi} \qquad u = 2\theta; \ du = 2 \, d\theta$$

$$= \frac{1}{2} \left[8\pi + 4 + \frac{1}{2}(2\pi) \right] - \frac{1}{2}[4]$$

$$= \frac{1}{2}(9\pi + 4) - 2 = \frac{9\pi}{2}$$

17.

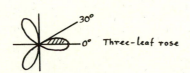

Since $r = 4\cos 3\theta = 0$ when $\theta = 0°$ and $\theta = 30°$, it follows that half a leaf is traced out in the interval $\theta = 0°$ to $\theta = 30°$. So integrating from $\theta = 0$ to $\theta = \frac{\pi}{6}$ yields one-sixth of the area:

171

$$\tfrac{1}{6}A \;=\; \tfrac{1}{2} \int_0^{\pi/6} (4 \cos 3\theta)^2 \, d\theta \;=\; 8 \int_0^{\pi/6} \cos^2 3\theta \, d\theta$$

$$=\; 8 \int_0^{\pi/6} \tfrac{1}{2}(1 + \cos 6\theta) \, d\theta \;=\; 4\Big(\theta + \tfrac{1}{6}\sin 6\theta\Big)\Big|_0^{\pi/6}$$

$$=\; 4\Big(\tfrac{\pi}{6}\Big) \;=\; \tfrac{2\pi}{3}$$

So $A = 6\Big(\tfrac{2\pi}{3}\Big) = 4\pi$

21.

If $r^2 = 4 \sin 2\theta$, then $r = 2\sqrt{\sin 2\theta}$, which is only the upper half of the lemniscate. Note that $r = 0$ when $\theta = 0°$ and $\theta = 90°$. The total area is therefore given by:

$$A \;=\; 2 \cdot \tfrac{1}{2} \int_0^{\pi/2} (2\sqrt{\sin 2\theta})^2 \, d\theta \;=\; \int_0^{\pi/2} 4 \sin 2\theta \, d\theta$$

$$=\; 4\Big(-\tfrac{1}{2}\cos 2\theta\Big)\Big|_0^{\pi/2} \;=\; -2 \cos 2\theta\big|_0^{\pi/2} \;=\; -2(-1 - 1) \;=\; 4$$

25.

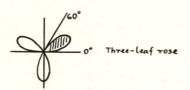

$0°$ Three-leaf rose

Since $r = 6 \sin 3\theta = 0$ when $\theta = 0°$ and $\theta = 60°$, it follows that one leaf is traced out in the interval $\theta = 0°$ to $\theta = 60°$. So integrating from $\theta = 0$ to $\theta = \tfrac{\pi}{3}$ yields one-third of the area:

$$\tfrac{1}{3}A \;=\; \tfrac{1}{2} \int_0^{\pi/3} (6 \sin 3\theta)^2 \, d\theta \;=\; \tfrac{1}{2} \int_0^{\pi/3} 36 \sin^2 3\theta \, d\theta$$

$$=\; 18 \int_0^{\pi/3} \tfrac{1}{2}(1 - \cos 6\theta) \, d\theta \;=\; 9 \int_0^{\pi/3} (1 - \cos 6\theta) \, d\theta$$

$$=\; 9\Big(\theta - \tfrac{1}{6}\sin 6\theta\Big)\Big|_0^{\pi/3} \;=\; 9\Big(\tfrac{\pi}{3}\Big) \;=\; 3\pi$$

So $A = 9\pi$

29.

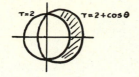

To determine where the curves intersect, we substitute $r = 2$ in $r = 2 + \cos\theta$ and solve for θ:

$$2 = 2 + \cos\theta$$

$$\cos\theta = 0$$

So $\theta = -90°, 90°$. We can therefore obtain the upper half of the shaded region by integrating from $\theta = 0$ to $\theta = \frac{\pi}{2}$:

$$\frac{1}{2}A = \frac{1}{2}\int_0^{\pi/2}(2 + \cos\theta)^2 d\theta - \frac{1}{2}\int_0^{\pi/2} 2^2 d\theta$$

$$A = \int_0^{\pi/2}(2 + \cos\theta)^2 d\theta - \int_0^{\pi/2} 4 d\theta$$

$$= \int_0^{\pi/2}(4 + 4\cos\theta + \cos^2\theta) d\theta - 4\theta\Big|_0^{\pi/2}$$

$$= \int_0^{\pi/2}\left[4 + 4\cos\theta + \frac{1}{2}(1 + \cos 2\theta)\right] d\theta - 4\left(\frac{\pi}{2}\right)$$

$$= 4\theta + 4\sin\theta + \frac{1}{2}\left(\theta + \frac{1}{2}\sin 2\theta\right)\Big|_0^{\pi/2} - 2\pi$$

$$= \left[2\pi + 4 + \frac{1}{2}\left(\frac{\pi}{2}\right)\right] - 0 - 2\pi$$

$$= 4 + \frac{\pi}{4} = \frac{16 + \pi}{4}$$

33.

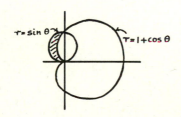

By inspection, we see that the curves intersect at $\theta = 90°$ and $\theta = 180°$.

$$A = \frac{1}{2} \int_{\pi/2}^{\pi} \sin^2\theta\, d\theta - \frac{1}{2} \int_{\pi/2}^{\pi} (1 + \cos\theta)^2 d\theta$$

$$= \frac{1}{2} \int_{\pi/2}^{\pi} \sin^2\theta\, d\theta - \frac{1}{2} \int_{\pi/2}^{\pi} (1 + 2\cos\theta + \cos^2\theta)\, d\theta$$

$$= \frac{1}{4} \int_{\pi/2}^{\pi} (1 - \cos 2\theta)\, d\theta - \frac{1}{2} \int_{\pi/2}^{\pi} \left[1 + 2\cos\theta + \frac{1}{2}(1 + \cos 2\theta)\right] d\theta$$

$$= \frac{1}{4}\left(\theta - \frac{1}{2}\sin 2\theta\right)\Big|_{\pi/2}^{\pi} - \frac{1}{2}\left[\theta + 2\sin\theta + \frac{1}{2}\left(\theta + \frac{1}{2}\sin 2\theta\right)\right]\Big|_{\pi/2}^{\pi}$$

$$= \frac{1}{4}\left(\pi - \frac{\pi}{2}\right) - \frac{1}{2}\left[\left(\pi + \frac{\pi}{2}\right) - \left(\frac{\pi}{2} + 2 + \frac{\pi}{4}\right)\right]$$

$$= \frac{1}{4}\left(\frac{\pi}{2}\right) - \frac{1}{2}\left(\frac{3\pi}{4} - 2\right) = 1 - \frac{\pi}{4}$$

35. Shaded area:

$$\frac{1}{2} \int_0^{\pi/2} 16(1 + \sin\theta)^2 d\theta$$

$$- \frac{1}{2} \int_0^{\pi/2} 64\,\sin^2\theta\, d\theta$$

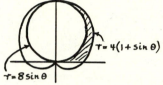

Area below polar axis (right side):

$$\frac{1}{2} \int_{-\pi/2}^{0} 16(1 + \sin\theta)^2 d\theta$$

37.

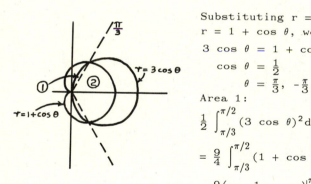

Substituting r = 3 cos θ in
r = 1 + cos θ, we get:

$$3\cos\theta = 1 + \cos\theta$$

$$\cos\theta = \frac{1}{2}$$

$$\theta = \frac{\pi}{3},\ -\frac{\pi}{3}$$

Area 1:

$$\frac{1}{2} \int_{\pi/3}^{\pi/2} (3\cos\theta)^2 d\theta = \frac{9}{2} \int_{\pi/3}^{\pi/2} \cos^2\theta\, d\theta$$

$$= \frac{9}{4} \int_{\pi/3}^{\pi/2} (1 + \cos 2\theta)\, d\theta$$

$$= \frac{9}{4}\left(\theta + \frac{1}{2}\sin 2\theta\right)\Big|_{\pi/3}^{\pi/2}$$

$$= \frac{9}{4}\left[\left(\frac{\pi}{2}\right) - \left(\frac{\pi}{3} + \frac{1}{2}\frac{\sqrt{3}}{2}\right)\right] = \frac{9}{4}\left(\frac{\pi}{2} - \frac{\pi}{3} - \frac{\sqrt{3}}{4}\right) = \frac{9}{4}\left(\frac{\pi}{6} - \frac{\sqrt{3}}{4}\right)$$

Area 2:

$$\frac{1}{2} \int_0^{\pi/3} (1 + \cos\theta)^2 d\theta = \frac{1}{2} \int_0^{\pi/3} (1 + 2\cos\theta + \cos^2\theta)\, d\theta$$

$$= \frac{1}{2} \int_0^{\pi/3} \left[1 + 2\cos\theta + \frac{1}{2}(1 + \cos 2\theta)\right] d\theta$$

$$= \frac{1}{2}\left[\theta + 2\sin\theta + \frac{1}{2}\left(\theta + \frac{1}{2}\sin 2\theta\right)\right]_0^{\pi/3}$$

$$= \frac{1}{2}\left(\frac{\pi}{3} + \frac{2\sqrt{3}}{2} + \frac{1}{2}\frac{\pi}{3} + \frac{1}{4}\frac{\sqrt{3}}{2}\right)$$

$$= \frac{1}{2}\left(\frac{\pi}{2} + \frac{9\sqrt{3}}{8}\right)$$

Total area $= 2(\text{Area } 1 + \text{Area } 2)$

$$= \frac{9}{2}\left(\frac{\pi}{6} - \frac{\sqrt{3}}{4}\right) + \left(\frac{\pi}{2} + \frac{9\sqrt{3}}{8}\right)$$

$$= \frac{9\pi}{12} - \frac{9\sqrt{3}}{8} + \frac{\pi}{2} + \frac{9\sqrt{3}}{8} = \frac{3\pi}{4} + \frac{2\pi}{4} = \frac{5\pi}{4}$$

REVIEW EXERCISES FOR CHAPTER 8

1.
$$x = 4\cos^2\theta$$
$$y = 4\sin\theta$$
$$\frac{x}{4} = \cos^2\theta$$
$$\frac{y^2}{16} = \sin^2\theta$$
$$\frac{x}{4} + \frac{y^2}{16} = \cos^2\theta + \sin^2\theta = 1$$
$$4x + y^2 = 16$$

5. $y = \ln\sec x$, $y' = \dfrac{\sec x \tan x}{\sec x} = \tan x$; $(y')^2 = \tan^2 x$

$1 + (y')^2 = 1 + \tan^2 x = \sec^2 x$

$$s = \int_0^{\pi/4}\sqrt{\sec^2 x}\,dx = \int_0^{\pi/4}\sec x\,dx = \ln\,|\sec x + \tan x\,\|_0^{\pi/4}$$

$$= \ln(\sqrt{2} + 1) - \ln 1 = \ln(1 + \sqrt{2}) \qquad (\text{since } \ln 1 = 0)$$

9. $r^2 = 4\tan\theta = 4\dfrac{\sin\theta}{\cos\theta} = 4\dfrac{r\sin\theta}{r\cos\theta}$

$$x^2 + y^2 = 4\frac{y}{x}$$

$$x(x^2 + y^2) = 4y$$

13. Four-leaf rose.

$r = 2$ when $\theta = 0°, 90°, 180°, 270°$
$r = 0$ when $\theta = 45°, 135°, 225°, 315°$

17. $A = \frac{1}{2} \int_0^{\pi/8} \tan^2 2\theta \, d\theta = \frac{1}{2} \int_0^{\pi/8} (\sec^2 2\theta - 1) \, d\theta$

$\qquad = \frac{1}{2}\left(\frac{1}{2} \tan 2\theta - \theta\right)\Big|_0^{\pi/8} = \frac{1}{2}\left(\frac{1}{2} - \frac{\pi}{8}\right) = \frac{1}{4} - \frac{\pi}{16}$

21.

$r = 0$ when $1 - 2 \cos \theta = 0$

$$\cos \theta = \frac{1}{2}$$

$$\theta = -\frac{\pi}{3}, \ \frac{\pi}{3}$$

$A = 2 \cdot \frac{1}{2} \int_0^{\pi/3} (1 - 2 \cos \theta)^2 \, d\theta = \int_0^{\pi/3} (1 - 4 \cos \theta + 4 \cos^2\theta) \, d\theta$

$\qquad = \int_0^{\pi/3} [1 - 4 \cos \theta + 2(1 + \cos 2\theta)] \, d\theta$

$\qquad = \theta - 4 \sin \theta + 2\theta + \sin 2\theta \Big|_0^{\pi/3}$

$\qquad = \frac{\pi}{3} - 4 \frac{\sqrt{3}}{2} + \frac{2\pi}{3} + \frac{\sqrt{3}}{2} = \pi - \frac{3}{2}\sqrt{3} = \frac{1}{2}(2\pi - 3\sqrt{3})$

25. If $x = f(\theta)$ and $y = g(\theta)$ are a set of parametric equations, then

$\dfrac{dy}{dx} = \dfrac{g'(\theta)}{f'(\theta)}$ by (8.2)

Now use the product rule:

$x = f(\theta) \cos \theta \qquad\qquad\qquad y = f(\theta) \sin \theta$

$\dfrac{dx}{d\theta} = f'(\theta) \cos \theta - f(\theta) \sin \theta \qquad \dfrac{dy}{d\theta} = f'(\theta) \sin \theta + f(\theta) \cos \theta$

$$\frac{dy}{dx} = \frac{f'(\theta) \sin \theta + f(\theta) \cos \theta}{f'(\theta) \cos \theta - f(\theta) \sin \theta}$$

CHAPTER 9   THREE-DIMENSIONAL SPACE; PARTIAL DERIVATIVES; MULTIPLE INTEGRALS

<u>Section 9.1</u>

1. The trace in the xy-plane is the circle $x^2 + y^2 = 4$.

In three-space, $x^2 + y^2 = 4$ represents a cylinder (note the missing z-variable) whose axis is the z-axis.

5. The trace in the yz-plane is the ellipse $y^2 + 4z^2 = 4$.

In three-space, $y^2 + 4z^2 = 4$ represents a cylinder (note the missing x-variable).   The cylinder extends along the x-axis: every cross-section parallel to the yz-plane is the ellipse $y^2 + 4z^2 = 4$.

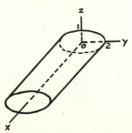

13. z = x is a line in the xz-plane. Because of the missing y-variable the plane z = x extends along the y-axis.

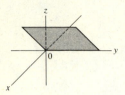

17. $9x^2 + 4y^2 + z^2 = 36$ (ellipsoid)

    1. Trace in xy-plane (z = 0): $9x^2 + 4y^2 = 36$

    2. Trace in xz-plane (y = 0): $9x^2 + z^2 = 36$

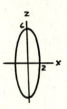

    3. Trace in yz-plane (x = 0): $4y^2 + z^2 = 36$

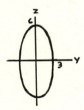
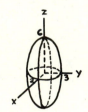

21. $z = 1 + 2x^2 + 4y^2$ (paraboloid)

1. Trace in xy-plane $(z = 0)$: $0 = 1 + 2x^2 + 4y^2$ or $2x^2 + 4y^2 = -1$
    (imaginary locus)

2. Trace in xz-plane $(y = 0)$: $z = 1 + 2x^2$

3. Trace in yz-plane $(x = 0)$: $z = 1 + 4y^2$

Cross-section: let $z = 2$ in the given equation. The resulting ellipse, $2x^2 + 4y^2 = 1$, is parallel to the xy-plane and two units above.

25. $-2x^2 - y^2 + z^2 = 6$

1. Trace in xy-plane $(z = 0)$: $-2x^2 - y^2 = 6$
    (imaginary locus)

2. Trace in xz-plane $(y = 0)$: $-2x^2 + z^2 = 6$

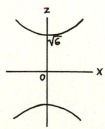

3. Trace in yz-plane $(x = 0)$: $-y^2 + z^2 = 6$

179

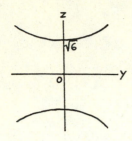

Cross-section: let z = 3 and z = -3 in the given equation.
In each case, the result is an ellipse:
$$-2x^2 - y^2 + 9 = 6 \qquad \text{or} \qquad 2x^2 + y^2 = 3$$

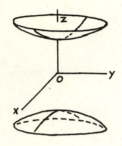

29. $9y^2 - 36x^2 - 16z^2 = 144$ (hyperboloid of two sheets)

   1. Trace in xy-plane (z = 0): $9y^2 - 36x^2 = 144$, or $y^2 - 4x^2 = 16$

   2. Trace in xz-plane (y = 0): $-36x^2 - 16z^2 = 144$
     (imaginary locus)

   3. Trace in yz-plane (x = 0): $9y^2 - 16z^2 = 144$

180

Cross-section: let $y = 8$ and $y = -8$ in the given equation. In each case, the resulting ellipse,

$$9(8)^2 - 36x^2 - 16z^2 = 144 \quad \text{or} \quad 9x^2 + 4z^2 = 108$$

is parallel to the xz-plane and 8 units to the right (respectively left).

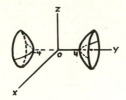

Section 9.2

1. $f(x,y) = 2x^2 + 5y^2 + 1$

    (a) $\frac{\partial f}{\partial x} = 4x + 0 = 4x$, since y is a constant

    (b) $\frac{\partial f}{\partial y} = 0 + 10y = 10y$, since x is a constant

5. $f(x,y) = x\sqrt{x + y} = x(x + y)^{1/2}$

    (a) Since x and $(x + y)^{1/2}$ are both functions of x, we need the product rule:

    $$\frac{\partial f}{\partial x} = x \frac{\partial}{\partial x}(x + y)^{1/2} + (x + y)^{1/2} \frac{\partial}{\partial x}(x)$$

    $$= x \cdot \frac{1}{2}(x + y)^{-1/2} + (x + y)^{1/2} \cdot 1$$

    $$= \frac{x}{2(x + y)^{1/2}} + \frac{(x + y)^{1/2}}{1} \cdot \frac{2(x + y)^{1/2}}{2(x + y)^{1/2}}$$

    $$= \frac{x + 2(x + y)^{1/2}(x + y)^{1/2}}{2(x + y)^{1/2}}$$

    $$= \frac{x + 2(x + y)}{2(x + y)^{1/2}} = \frac{3x + 2y}{2\sqrt{x + y}}$$

    (b) This time x is a constant coefficient:

    $$\frac{\partial f}{\partial y} = x \frac{\partial}{\partial y}(x + y)^{1/2} = x \cdot \frac{1}{2}(x + y)^{-1/2} = \frac{x}{2\sqrt{x + y}}$$

9. $f(x,y) = \frac{\cos x^2 y}{y}$

    (a) $f(x,y) = \frac{1}{y}\cos x^2 y$. Since y is a constant coefficient, we get

    $$\frac{\partial f}{\partial x} = \frac{1}{y} \frac{\partial}{\partial x}\cos x^2 y = \frac{1}{y}(-\sin x^2 y) \frac{\partial}{\partial x}(x^2 y)$$

181

$$= \tfrac{1}{y}(-\sin\ x^2 y)(2xy) = -2x\ \sin\ x^2 y$$

(b) Since y and cos $x^2 y$ are both functions of y, we need the quotient rule:

$$\frac{\partial f}{\partial y} = \frac{y\frac{\partial}{\partial y}\ \cos\ x^2 y\ -\ \cos\ x^2 y\ \frac{\partial}{\partial y}(y)}{y^2}$$

$$= \frac{y(-\sin\ x^2 y)\frac{\partial}{\partial y}(x^2 y)\ -\ \cos\ x^2 y\ \cdot\ 1}{y^2}$$

$$= \frac{y(-\sin\ x^2 y)(x^2)\ -\ \cos\ x^2 y}{y^2}$$

$$= \frac{-x^2 y\ \sin\ x^2 y\ -\ \cos\ x^2 y}{y^2}$$

13. $f(x,y) = 5x - 2y - 3$

(a) $\frac{\partial f}{\partial x} = 5$ (b) $\frac{\partial f}{\partial y} = -2$

(c) $\frac{\partial^2 f}{\partial x^2} = 0$ (d) $\frac{\partial^2 f}{\partial y^2} = 0$

(e) $\frac{\partial^2 f}{\partial x \partial y} = 0$

17. $f(x,y) = \sin(2x + y)$

(a) $\frac{\partial f}{\partial x} = 2\ \cos(2x + y)$ (b) $\frac{\partial f}{\partial y} = \cos(2x + y)$

(c) $\frac{\partial^2 f}{\partial x^2} = \frac{\partial}{\partial x}[2\ \cos(2x + y)] = -4\ \sin(2x + y)$

(d) $\frac{\partial^2 f}{\partial y^2} = \frac{\partial}{\partial y}[\cos(2x + y)] = -\sin(2x + y)$

(e) $\frac{\partial^2 f}{\partial x \partial y} = \frac{\partial}{\partial x}\left(\frac{\partial f}{\partial y}\right) = \frac{\partial}{\partial x}[\cos(2x + y)] = -2\ \sin(2x + y)$

or

$$\frac{\partial}{\partial y}[2\ \cos(2x + y)] = -2\ \sin(2x + y)$$

21. $f(x,y) = \text{Arctan}\ \frac{y}{x}$

(a) $\frac{\partial f}{\partial x} = \frac{1}{1 + \left(\frac{y}{x}\right)^2}\left(-\frac{y}{x^2}\right) = -\frac{y}{x^2 + y^2}$

(b) $\frac{\partial f}{\partial y} = \frac{1}{1 + \left(\frac{y}{x}\right)^2}\left(\frac{1}{x}\right) = \frac{1}{1 + \frac{y^2}{x^2}} \cdot \frac{1}{x} \cdot \frac{x}{x} = \frac{x}{x^2 + y^2}$

(c) $\dfrac{\partial^2 f}{\partial x^2} = \dfrac{\partial}{\partial x}\left[-y(x^2 + y^2)^{-1}\right] = y(x^2 + y^2)^{-2}(2x) = \dfrac{2xy}{(x^2 + y^2)^2}$

(d) $\dfrac{\partial^2 f}{\partial y^2} = \dfrac{\partial}{\partial y}\left[x(x^2 + y^2)^{-1}\right] = -x(x^2 + y^2)^{-2}(2y) = -\dfrac{2xy}{(x^2 + y^2)^2}$

(e) $\dfrac{\partial^2 f}{\partial x \partial y} = \dfrac{\partial}{\partial x}\left(\dfrac{\partial f}{\partial y}\right) = \dfrac{\partial}{\partial x}\ \dfrac{x}{x^2 + y^2}$

$\qquad = \dfrac{(x^2 + y^2)\cdot 1 - x\cdot 2x}{(x^2 + y^2)^2}$ quotient rule

$\qquad = \dfrac{-x^2 + y^2}{(x^2 + y^2)^2}$

25. (a) $\dfrac{\partial}{\partial x}(ye^x) = ye^x\Big|_{(1,-1,-e)} = (-1)e = -e$

    (b) $\dfrac{\partial}{\partial y}(ye^x) = e^x\Big|_{(1,-1,-e)} = e$

29. $Z = \dfrac{RX}{R + X}$. By the quotient rule

$\dfrac{\partial Z}{\partial R} = \dfrac{(R + X)(X) - RX(1)}{(R + X)^2} = \dfrac{RX + X^2 - RX}{(R + X)^2} = \dfrac{X^2}{(R + X)^2}$

If $R = 8.0$ and $X = 5.0$,

$\dfrac{\partial Z}{\partial R} = \dfrac{(5.0)^2}{(8.0 + 5.0)^2} = 0.15$

33. $g_m = \dfrac{\partial}{\partial v_g}\left[0.50(v_g + 0.1v_p)^{4/3}\right] = 0.50\left(\dfrac{4}{3}\right)(v_g + 0.1v_p)^{1/3}(1)$

$\qquad\qquad\qquad\qquad\qquad = 0.67(v_g + 0.1v_p)^{1/3}\Omega^{-1}$

37. $y = A\cos\omega\left(t - \dfrac{x}{v}\right)$

$\dfrac{\partial y}{\partial x} = -A\sin\omega\left(t - \dfrac{x}{v}\right)\dfrac{\partial}{\partial x}\ \omega\left(t - \dfrac{x}{v}\right) = \left[-A\sin\omega\left(t - \dfrac{x}{v}\right)\right]\left(-\dfrac{\omega}{v}\right)$

$\qquad = \left(\dfrac{\omega}{v}\right)A\sin\omega\left(t - \dfrac{x}{v}\right)$

$\dfrac{\partial^2 y}{\partial x^2} = \left[\left(\dfrac{\omega}{v}\right)A\cos\omega\left(t - \dfrac{x}{v}\right)\right]\left(-\dfrac{\omega}{v}\right) = \left(-\dfrac{\omega^2}{v^2}\right)A\cos\omega\left(t - \dfrac{x}{v}\right)$

$\qquad = \dfrac{1}{v^2}\dfrac{\partial^2 y}{\partial t^2}$ since $\dfrac{\partial^2 y}{\partial t^2} = -\omega^2\left[A\cos\omega\left(t - \dfrac{x}{v}\right)\right]$

Section 9.3

1. $P = \dfrac{2L}{V^2} = 2LV^{-2}$

$dP = \dfrac{\partial}{\partial L}(2LV^{-2})dL + \dfrac{\partial}{\partial V}(2LV^{-2})dV$

$\quad = 2V^{-2}dL + (2L)(-2V^{-3})dV$

$$= \frac{2}{V^2}dL - \frac{4L}{V^3}dV$$

5. $M = \frac{\sin \theta_1}{\sin \theta_2} = \sin \theta_1 \csc \theta_2$

$dM = \frac{\partial}{\partial \theta_1}(\sin \theta_1 \csc \theta_2)d\theta_1 + \frac{\partial}{\partial \theta_2}(\sin \theta_1 \csc \theta_2)d\theta_2$

$= \cos \theta_1 \csc \theta_2 d\theta_1 + (\sin \theta_1)(-\csc \theta_2 \cot \theta_2)d\theta_2$

$= \frac{\cos \theta_1}{\sin \theta_2} d\theta_1 - \frac{\sin \theta_1}{\sin \theta_2 \tan \theta_2} d\theta_2$

9. $V = \frac{L}{P^2} = LP^{-2}$

Differential:
$dV = P^{-2}dL + L(-2P^{-3})dP$

$dV = \frac{1}{P^2}dL - \frac{2L}{P^3}dP$

$$L = 8.0, \quad dL = \pm 0.5$$
$$P = 3.0, \quad dP = \pm 0.2$$

Approximate maximum error: either

$dV = \frac{1}{(3.0)^2}(+0.5) - \frac{2(8.0)}{(3.0)^3}(-0.2)$

or

$dV = \frac{1}{(3.0)^2}(-0.5) - \frac{2(8.0)}{(3.0)^3}(+0.2)$

Thus

$dV = \pm 0.17$

13. $i = 1 - e^{-R/L}$

Differential:

$di = \frac{\partial}{\partial L}\left(1 - e^{-R/L}\right)dL + \frac{\partial}{\partial R}\left(1 - e^{-R/L}\right)dR$

$= -e^{-R/L}\left(\frac{R}{L^2}\right)dL + \left(-e^{-R/L}\right)\left(-\frac{1}{L}\right)dR$

$di = e^{-R/L}\left[-\frac{R}{L^2}dL + \frac{1}{L}dR\right]$

$R = 1.2, \quad dR = \pm 0.05$
$L = 0.70, \quad dL = \pm 0.01$

Approximate maximum error: either

$di = e^{-1.2/0.70}\left[-\frac{1.2}{(0.70)^2}(-0.01) + \frac{1}{0.70}(0.05)\right]$

or

$$di = e^{-1.2/0.70}\left[-\frac{1.2}{(0.70)^2}(0.01) + \frac{1}{0.70}(-0.05)\right]$$

Thus

$$di = \pm 0.017 \text{ A}$$

Approximate maximum percentage error:

$$\frac{di}{i} \times 100 = \frac{0.017}{1 - e^{-1.2/0.70}} \times 100 = 2.1\%$$

17. $V = \frac{1}{3}\pi r^2 h$.

Given: $\frac{dr}{dt} = 1.0 \frac{\text{cm}}{\text{min}}$, $\frac{dh}{dt} = 1.0 \frac{\text{cm}}{\text{min}}$

Find: $\frac{dV}{dt}$ when $r = 10$ cm and $h = 20$ cm

$$\frac{dV}{dt} = \frac{\partial V}{\partial r}\frac{dr}{dt} + \frac{\partial V}{\partial h}\frac{dh}{dt}$$

$$= \left(\frac{2}{3}\pi rh\right)\frac{dr}{dt} + \left(\frac{1}{3}\pi r^2\right)\frac{dh}{dt}$$

$$= \left(\frac{2}{3}\pi rh\right)(1.0) + \left(\frac{1}{3}\pi r^2\right)(1.0)\Big|_{r=10,h=20}$$

$$= \frac{500\pi}{3} = 520 \text{ cm}^3/\text{min}$$

21. $z = 3x^2 + 2y^2 + 4x - 4y - 1$

Critical points:

$\frac{\partial z}{\partial x} = 6x + 4 = 0$ or $x = -\frac{2}{3}$

$\frac{\partial z}{\partial y} = 4y - 4 = 0$ or $y = 1$

The critical point is at $\left(-\frac{2}{3}, 1\right)$

$\frac{\partial^2 z}{\partial x^2} = 6$, $\frac{\partial^2 z}{\partial y^2} = 4$, $\frac{\partial^2 z}{\partial x \partial y} = 0$

Thus $A = 6 \cdot 4 - 0^2 = 24 > 0$

Since $A > 0$ and $\frac{\partial^2 z}{\partial x^2} > 0$, $f(x,y)$ has a minimum at $\left(-\frac{2}{3}, 1\right)$.

Also, $z = 3\left(-\frac{2}{3}\right)^2 + 2(1)^2 + 4\left(-\frac{2}{3}\right) - 4(1) - 1$

$$= \frac{4}{3} + 2 - \frac{8}{3} - 4 - 1 = -3 - \frac{4}{3} = -\frac{13}{3}$$

25. $z = y^2 - x^2 - 2xy - 4y$

Critical points:

$$\frac{\partial z}{\partial x} = -2x - 2y = 0$$

$$\frac{\partial z}{\partial y} = \underline{2y - 2x - 4 = 0}$$

$$-4x - 4 = 0 \qquad \text{adding}$$

$$x = -1$$

$$y = 1$$

The critical point is at $(-1,1)$; $z = 1^2 - (-1)^2 - 2(-1)(1) - 4(1) = -2$

$$\frac{\partial^2 z}{\partial x^2} = -2, \ \frac{\partial^2 z}{\partial y^2} = 2, \ \frac{\partial^2 z}{\partial x \partial y} = -2$$

Thus

$$A = (-2)(2) - (-2)^2 = -8 < 0$$

Since $A < 0$, the point $(-1,1,-2)$ is a saddle point.

29. $z = 3x^3 - xy^2 + x$

Critical points:

$$\frac{\partial z}{\partial x} = 9x^2 - y^2 + 1 = 0$$

$$\frac{\partial z}{\partial y} = -2xy = 0, \text{ which implies that } x = 0 \text{ or } y = 0$$

Substituting $y = 0$ in the first equation, we get

$$9x^2 + 1 = 0$$

which has no real roots. If we substitute $x = 0$ in the first equation, we get

$$-y^2 + 1 = 0, \qquad \text{whence} \qquad y = \pm 1$$

so the critical points are at $(0,1)$ and $(0,-1)$.

$$\frac{\partial^2 z}{\partial x^2} = 18x, \ \frac{\partial^2 z}{\partial y^2} = -2x, \ \frac{\partial^2 z}{\partial x \partial y} = -2y$$

So

$$A = (18x)(-2x) - (-2y)^2$$

or

$$A = -36x^2 - 4y^2$$

For both critical points,

$$A = -4 < 0$$

We conclude that $f(x,y)$ has saddle points at $(0,1)$ and $(0,-1)$.

33.   Let x and y be the first two numbers; then $60 - x - y$ is the third.
The quantity to be maximized is the product P:

$$P = xy(60 - x - y)$$

$$P = 60xy - x^2y - xy^2$$

$$\frac{\partial P}{\partial x} = 60y - 2xy - y^2 = 0$$

$$\frac{\partial P}{\partial y} = 60x - x^2 - 2xy = 0$$

$$-2xy = x^2 - 60x \qquad \text{second equation}$$

$$-2xy = x^2 - 60x \qquad \text{second equation}$$

$$y = -\tfrac{1}{2}x + 30$$

Substituting in first equation:

$$60\left(-\tfrac{1}{2}x + 30\right) - 2x\left(-\tfrac{1}{2}x + 30\right) - \left(-\tfrac{1}{2}x + 30\right)^2 = 0$$

$$-30x + 1800 + x^2 - 60x - \tfrac{1}{4}x^2 + 30x - 900 = 0$$

$$\tfrac{3}{4}x^2 - 60x + 900 = 0$$

$$x^2 - 80x + 1200 = 0$$

$$(x - 20)(x - 60) = 0$$

$$x = 20$$

$$y = 20$$

third number $= 60 - 20 - 20 = 20$

37.
$$f(x,y) = 0$$

$$\frac{\partial f}{\partial x}\, dx + \frac{\partial f}{\partial y}\, dy = 0$$

$$\frac{\partial f}{\partial x} + \frac{\partial f}{\partial y}\frac{dy}{dx} = 0$$

$$\frac{dy}{dx} = -\frac{\frac{\partial f}{\partial x}}{\frac{\partial f}{\partial y}}$$

41. Let
$$f = 2x^5 + 3x^4y - 4x^3y^2 + 7x^2y^2 + 1$$

$$\frac{\partial f}{\partial x} = 10x^4 + 12x^3y - 12x^2y^2 + 14xy^2$$

$$\frac{\partial f}{\partial y} = 3x^4 - 8x^3y + 14x^2y$$

By Exercise 37,
$$\frac{dy}{dx} = -\frac{\partial f/\partial x}{\partial f/\partial y} = -\frac{10x^4 + 12x^3y - 12x^2y^2 + 14xy^2}{3x^4 - 8x^3y + 14x^2y}$$

45. Let
$$f = 7x^4y^8 + 16x^3y^5 + 25x^2 - 7y^2 + 2 = 0$$

$$\frac{\partial f}{\partial x} = 28x^3y^8 + 48x^2y^5 + 50x$$

$$\frac{\partial f}{\partial y} = 56x^4y^7 + 80x^3y^4 - 14y$$

By Exercise 37,
$$\frac{dy}{dx} = -\frac{\partial f/\partial x}{\partial f/\partial y} = -\frac{28x^3y^8 + 48x^2y^5 + 50x}{56x^4y^7 + 80x^3y^4 - 14y}$$

49. $f = x \tan y - 3y^2 + 5$

$\dfrac{\partial f}{\partial x} = \tan y$

$\dfrac{\partial f}{\partial y} = x \sec^2 y - 6y$

$$\dfrac{dy}{dx} = -\dfrac{\dfrac{\partial f}{\partial x}}{\dfrac{\partial f}{\partial y}} = -\dfrac{\tan y}{x \sec^2 y - 6y} = \dfrac{\tan y}{6y - x \sec^2 y}$$

53. $f = y^2 e^x - e^{-xy} + 3$

$\dfrac{\partial f}{\partial x} = y^2 e^x - e^{-xy}(-y) = y^2 e^x + y e^{-xy}$

$\dfrac{\partial f}{\partial y} = 2y e^x - e^{-xy}(-x) = 2y e^x + x e^{-xy}$

$$\dfrac{dy}{dx} = -\dfrac{\dfrac{\partial f}{\partial x}}{\dfrac{\partial f}{\partial y}} = -\dfrac{y^2 e^x + y e^{-xy}}{2y e^x + x e^{-xy}}$$

Section 9.4

1.

x_i	y_i	x_i^2	$x_i y_i$
2	14	4	28
4	36	16	144
6	53	36	318
8	78	64	624
10	92	100	920

Totals   30   273   220   2034 $n = 5$

$D = \begin{vmatrix} 30 & 220 \\ 5 & 30 \end{vmatrix} = (30)(30) - (5)(220) = -200$

$A = \begin{vmatrix} 30 & 2034 \\ 5 & 273 \end{vmatrix} = (30)(273) - (5)(2034) = -1980$

$B = \begin{vmatrix} 2034 & 220 \\ 273 & 30 \end{vmatrix} = (2034)(30) - (273)(220) = 960$

$a = \dfrac{A}{D} = \dfrac{-1980}{-200} = 9.9, \quad b = \dfrac{B}{D} = \dfrac{960}{-200} = -4.8$

Since $y = ax + b$, we get $y = 9.9x - 4.8$

5. We let $T_1' = T_1^2 = (20)^2$, $T_2' = T_2^2 = (30)^2$, and so on:

T_i'		R_i	$(T_i')^2$	$T_i'R_i$
$20^2 =$	400	7	160000	2800
$30^2 =$	900	8	810000	7200
$40^2 =$	1600	11	2560000	17600
$50^2 =$	2500	16	6250000	40000
$60^2 =$	3600	22	12960000	79200
$70^2 =$	4900	27	24010000	132300

| Totals | 13900 | 91 | 46750000 | 279100 | n = 6 |

$$D = \begin{vmatrix} 13900 & 46750000 \\ 6 & 13900 \end{vmatrix} = -87290000$$

$$A = \begin{vmatrix} 13900 & 279100 \\ 6 & 91 \end{vmatrix} = -409700$$

$$B = \begin{vmatrix} 279100 & 46750000 \\ 91 & 13900 \end{vmatrix} = -3.7476 \times 10^8$$

$$a = \frac{A}{D} = \frac{-409700}{-87290000} = 0.0047, \quad b = \frac{B}{D} = 4.2933$$

Thus

$$R = 0.0047T^2 + 4.2933$$

Section 9.5

1. $\displaystyle\int_0^1\left(\int_0^x x^2y^2\,dy\right)dx = \int_0^1 x^2\,\frac{y^3}{3}\Big|_0^x\,dx = \int_0^1 x^2\left(\frac{x^3}{3} - 0\right)dx$

$\displaystyle = \int_0^1 \tfrac{1}{3}x^5\,dx = \tfrac{1}{3}\,\frac{x^6}{6}\Big|_0^1 = \frac{1}{18}$

5. $\displaystyle\int_0^2\int_0^{\sqrt{y-1}} xy\,dx\,dy = \int_0^2 y\cdot\tfrac{1}{2}x^2\Big|_0^{\sqrt{y-1}}\,dy$

$\displaystyle = \int_0^2 y\cdot\tfrac{1}{2}\left(\sqrt{y-1}\right)^2 dy = \tfrac{1}{2}\int_0^2 y(y-1)\,dy = \tfrac{1}{2}\int_0^2 (y^2 - y)\,dy$

$\displaystyle = \tfrac{1}{2}\left(\tfrac{1}{3}y^3 - \tfrac{1}{2}y^2\right)\Big|_0^2 = \tfrac{1}{2}\left(\tfrac{8}{3} - 2\right) = \tfrac{1}{3}$

9. $\displaystyle\int_0^{\sqrt{\pi/6}}\left(\int_0^x \cos x^2\,dy\right)dx = \int_0^{\sqrt{\pi/6}} (\cos x^2)y\big|_0^x\,dx = \int_0^{\sqrt{\pi/6}} (\cos x^2)x\,dx$

$\displaystyle \qquad\qquad = \tfrac{1}{2}\int_0^{\sqrt{\pi/6}} \cos x^2(2x)\,dx \qquad\qquad u = x^2;\ du = 2x\,dx$

189

$$= \frac{1}{2} \sin x^2 \Big|_0^{\sqrt{\pi/6}} = \frac{1}{2} \sin \frac{\pi}{6} = \frac{1}{4}$$

13. $$\int_0^3 \left(\int_0^x e^x \, dy \right) dx = \int_0^3 y e^x \Big|_0^x \, dx = \int_0^3 x e^x \, dx$$

$$u = x \qquad dv = e^x \, dx$$
$$du = dx \qquad v = e^x$$

$$x e^x \Big|_0^3 - \int_0^3 e^x \, dx = x e^x \Big|_0^3 - e^x \Big|_0^3 = 3e^3 - (e^3 - 1) = 2e^3 + 1$$

17.

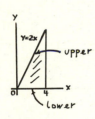

$$A = \int_0^2 \int_0^{(2-x)/2} dy \, dx = \int_0^2 y \Big|_0^{(2-x)/2} \, dx$$

$$= \int_0^2 \frac{1}{2}(2 - x) \, dx = x - \frac{1}{4} x^2 \Big|_0^2 = 1$$

$$A = \int_0^1 \int_0^{2-2y} dx \, dy = \int_0^1 x \Big|_0^{2-2y} \, dy$$

$$= \int_0^1 (2 - 2y) \, dy = 2y - y^2 \Big|_0^1 = 2 - 1 = 1$$

21.

$$A = \int_0^4 \int_0^{2x} dy \, dx = \int_0^4 y \Big|_0^{2x} \, dx = \int_0^4 2x \, dx$$

$$= x^2 \Big|_0^4 = 16$$

$$A = \int_0^8 \int_{(1/2)y}^4 dx \, dy = \int_0^8 x \Big|_{(1/2)y}^4 \, dy = \int_0^8 \left(4 - \frac{1}{2} y \right) dy$$

$$= 4y - \frac{1}{4} y^2 \Big|_0^8 = 32 - 16 = 16$$

25.

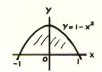

$$A = \int_0^1 \int_x^{\sqrt{x}} dy\,dx = \int_0^1 y\big|_x^{\sqrt{x}}\ dx = \int_0^1 (\sqrt{x}\ -\ x)\,dx$$

$$= \tfrac{2}{3}x^{3/2}\ -\ \tfrac{x^2}{2}\Big|_0^1 = \tfrac{2}{3}\ -\ \tfrac{1}{2} = \tfrac{1}{6}$$

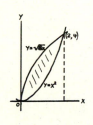

$$A = \int_0^1 \int_{y^2}^y dx\,dy = \int_0^1 x\big|_{y^2}^y\ dy = \int_0^1 (y\ -\ y^2)\,dy$$

$$= \tfrac{1}{2}y^2\ -\ \tfrac{1}{3}y^3\big|_0^1 = \tfrac{1}{2}\ -\ \tfrac{1}{3} = \tfrac{1}{6}$$

29.

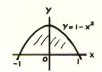

$$A = \int_{-1}^1 \int_0^{1-x^2} dy\,dx = \int_{-1}^1 y\big|_0^{1-x^2}\ dx = \int_{-1}^1 (1\ -\ x^2)\,dx$$

$$= x\ -\ \tfrac{1}{3}x^3\Big|_{-1}^1 = \left(1\ -\ \tfrac{1}{3}\right)\ -\ \left(-1\ +\ \tfrac{1}{3}\right) = \tfrac{4}{3}$$

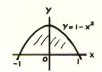

$$A = \int_0^1 \int_{-\sqrt{1-y}}^{\sqrt{1-y}} dx\,dy = \int_0^1 x\big|_{-\sqrt{1-y}}^{\sqrt{1-y}}\ dy$$

$$= \int_0^1 (\sqrt{1\ -\ y}\ +\ \sqrt{1\ -\ y})\,dy$$

$$= 2\int_0^1 \sqrt{1\ -\ y}\,dy \qquad\qquad \begin{array}{l} u\ =\ 1-y; \\ du\ =\ -dy \end{array}$$

$$= -2\int_0^1 (1\ -\ y)^{1/2}(-dy) = -2\cdot\tfrac{2}{3}(1\ -\ y)^{3/2}\big|_0^1 = 0 + \tfrac{4}{3}\cdot 1^{3/2} = \tfrac{4}{3}$$

33.

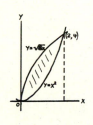

$$y\ =\ x^2$$

$$\underline{y^2\ =\ 8x}$$

$$y^2\ =\ x^4$$

$$\underline{y^2\ =\ 8x}$$

$$0\ =\ x^4\ -\ 8x \qquad \text{(subtracting)}$$

$$x(x^3\ -\ 8)\ =\ 0$$

$$x\ =\ 0,2$$

$$A = \int_0^2 \int_{x^2}^{\sqrt{8x}} dy\,dx = \int_0^2 y\big|_{x^2}^{\sqrt{8x}}\ dx = \int_0^2 (\sqrt{8x} - x^2)\,dx$$

$$= \int_0^2 \left(2\sqrt{2}\,x^{1/2} - x^2\right)dx = 2\sqrt{2}\cdot\tfrac{2}{3}x^{3/2} - \tfrac{1}{3}x^3\Big|_0^2 = 2\sqrt{2}\cdot\tfrac{2}{3}\cdot 2^{3/2} - \tfrac{1}{3}\cdot 8$$

$$= \tfrac{4}{3}\cdot 2^{1/2}\cdot 2^{3/2} - \tfrac{8}{3}$$

$$= \tfrac{4}{3}\cdot 2^2 - \tfrac{8}{3} = \tfrac{16}{3} - \tfrac{8}{3} = \tfrac{8}{3}$$

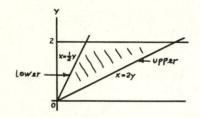

$$A = \int_0^4 \int_{y^2/8}^{\sqrt{y}} dx\,dy = \int_0^4 x\big|_{y^2/8}^{\sqrt{y}}\ dy = \int_0^4 \left(\sqrt{y} - \tfrac{1}{8}y^2\right)dy$$

$$= \left(\tfrac{2}{3}y^{3/2} - \tfrac{1}{24}y^3\right)\Big|_0^4 = \tfrac{2}{3}\cdot 4^{3/2} - \tfrac{1}{24}\cdot 64$$

$$= \tfrac{2}{3}\cdot 8 - \tfrac{8}{3} = \tfrac{16}{3} - \tfrac{8}{3} = \tfrac{8}{3}$$

37.

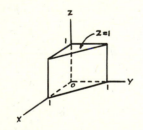

$$A = \int_0^2 \int_{y/2}^{2y} dx\,dy = \int_0^2 x\big|_{y/2}^{2y}\ dy = \int_0^2 \left(2y - \tfrac{1}{2}y\right)dy = \int_0^2 \tfrac{3}{2}y\,dy$$

$$= \tfrac{3}{2}\cdot\tfrac{y^2}{2}\Big|_0^2 = 3$$

Section 9.6

1.

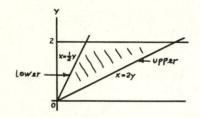

$$V = \int_0^1 \int_0^{1-x} 1\,dy\,dx = \int_0^1 y\big|_0^{1-x}\ dx = \int_0^1 (1 - x)\,dx = \tfrac{1}{2}$$

5.

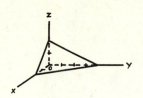

If z = 0, we get y = 4 - 2x, which is the trace in the xy-plane. Thus y = 0 is the lower function and y = 4 - 2x is the upper function. The integrand is $z = \frac{1}{2}(4 - 2x - y)$.

$$V = \int_0^2 \int_0^{4-2x} \left(\tfrac{1}{2}\right)(4 - 2x - y)\,dy\,dx = \tfrac{1}{2}\int_0^2 \left(4y - 2xy - \tfrac{1}{2}y^2\right)\Big|_0^{4-2x}\,dx$$

$$= \tfrac{1}{2}\int_0^2 \left[4(4 - 2x) - 2x(4 - 2x) - \tfrac{1}{2}(4 - 2x)^2\right]dx$$

$$= \tfrac{1}{2}\int_0^2 (16 - 8x - 8x + 4x^2 - 8 + 8x - 2x^2)\,dx$$

$$= \tfrac{1}{2}\int_0^2 (2x^2 - 8x + 8)\,dx = \tfrac{1}{2}\left(\tfrac{2}{3}x^3 - 4x^2 + 8x\right)\Big|_0^2$$

$$= \tfrac{1}{2}\left(\tfrac{16}{3} - 16 + 16\right) = \tfrac{8}{3}$$

9. From $x^2 + y^2 = 9$, we get $y = \sqrt{9 - x^2}$.

 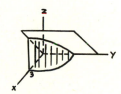

$$V = \int_0^3 \int_0^{\sqrt{9-x^2}} x\,dy\,dx = \int_0^3 xy\big|_0^{\sqrt{9-x^2}}\,dx$$

$$= \int_0^3 x\sqrt{9 - x^2}\,dx = -\tfrac{1}{2}\int_0^3 (9 - x^2)^{1/2}(-2x)\,dx \qquad u = 9 - x^2;$$
$$du = -2x\,dx$$

$$= -\tfrac{1}{2}\cdot\tfrac{2}{3}(9 - x^2)^{3/2}\Big|_0^3 = 0 + \tfrac{1}{3}(9)^{3/2} = 9$$

13.

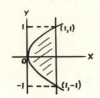

Since the paraboloid lies above the region, a two-dimensional figure is sufficient for obtaining the limits of integration. Since x is given as a function of y, it is convenient to use (9.14) with

$$h_1(y) = y^2$$

and

$$h_2(y) = 1.$$

Thus

$$V = \int_{-1}^{1} \int_{y^2}^{1} (x^2 + 3y^2)\,dx\,dy = \int_{-1}^{1} \left(\tfrac{1}{3}x^3 + 3xy^2\right)\Big|_{y^2}^{1}\,dy$$

$$= \int_{-1}^{1} \left[\left(\tfrac{1}{3} + 3y^2\right) - \left(\tfrac{1}{3}y^6 + 3y^4\right)\right]dy$$

$$= \int_{-1}^{1} \left(\tfrac{1}{3} + 3y^2 - \tfrac{1}{3}y^6 - 3y^4\right)dy$$

$$= \tfrac{1}{3}y + y^3 - \tfrac{1}{21}y^7 - \tfrac{3}{5}y^5\Big|_{-1}^{1} = \left(\tfrac{1}{3} + 1 - \tfrac{1}{21} - \tfrac{3}{5}\right) - \left(-\tfrac{1}{3} - 1 + \tfrac{1}{21} + \tfrac{3}{5}\right)$$

$$= 2\left(\tfrac{1}{3} + 1 - \tfrac{1}{21} - \tfrac{3}{5}\right) = (2)\,\frac{35 + 105 - 5 - 63}{105}$$

$$= \frac{144}{105} = \frac{48}{35}$$

17.

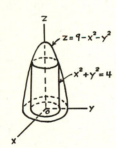

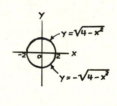

$$V = \int_{-2}^{2} \int_{-\sqrt{4-x^2}}^{\sqrt{4-x^2}} (9 - x^2 - y^2)\,dy\,dx$$

19.

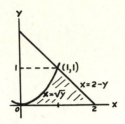

By (9.14) with $h_1(y) = \sqrt{y}$ and $h_2(y) = 2 - y$, we get:

$$V = \int_0^1 \int_{\sqrt{y}}^{2-y} \sqrt{8 - 2x^2 - y^2}\, dx\, dy$$

21.

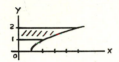

By (9.14) with $h_1(y) = 0$ and $h_2(y) = y^2 + 1$, we get

$$V = \int_1^2 \int_0^{y^2+1} xy\, dx\, dy$$

25.

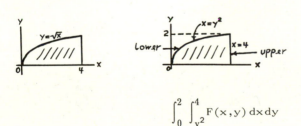

$$\int_0^2 \int_{y^2}^4 F(x,y)\, dx\, dy$$

29.

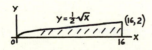

From $x = 4y^2$, we get $y = \frac{1}{2}\sqrt{x}$. So by (9.13) with $g_1(x) = 0$ and and $g_2(x) = \frac{1}{2}\sqrt{x}$, we get:

195

$$V = \int_0^{16} \int_0^{\sqrt{x}/2} F(x,y)\,dy\,dx$$

33.

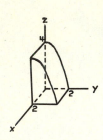

$$V = \int_0^2 \int_0^2 (4 - y^2)\,dy\,dx$$

$$= \int_0^2 \left(4y - \tfrac{1}{3}y^3\big|_0^2\right)\,dx$$

$$= \int_0^2 \left(8 - \tfrac{8}{3}\right)dx$$

$$= \tfrac{16}{3} \int_0^2 dx = \tfrac{16}{3}(2) = \tfrac{32}{3}$$

Section 9.7

1. Mass of typical element: $\tfrac{1}{4}x\,dy\,dx$

$$m = \int_0^4 \int_0^{x/2} \tfrac{1}{4}x\,dy\,dx$$

$$= \int_0^4 \tfrac{1}{4}xy\Big|_0^{x/2}\,dx$$

$$= \int_0^4 \tfrac{1}{4}x\left(\tfrac{1}{2}x\right)dx = \tfrac{8}{3}$$

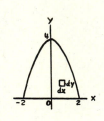

5. Mass of typical element: $x^2\,dy\,dx$

$$m = \int_{-2}^2 \int_0^{4-x^2} x^2\,dy\,dx = \int_{-2}^2 x^2 y\big|_0^{4-x^2}\,dx$$

$$= \int_{-2}^2 x^2(4 - x^2)\,dx = \tfrac{4}{3}x^3 - \tfrac{1}{5}x^5\big|_{-2}^2$$

$$= \left(\tfrac{32}{3} - \tfrac{32}{5}\right) - \left(-\tfrac{32}{3} + \tfrac{32}{5}\right)$$

$$= 2\left(\tfrac{32}{3} - \tfrac{32}{5}\right) = 64\left(\tfrac{1}{3} - \tfrac{1}{5}\right)$$

$$= 64\left(\tfrac{2}{15}\right) = \tfrac{128}{15}$$

9.

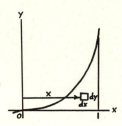

Moment of inertia of typical element: $\rho x^2 dy\,dx$

$$I_y = \rho \int_0^1 \int_0^{x^3} x^2\,dy\,dx = \rho \int_0^1 x^2 y\big|_0^{x^3}\,dx = \rho \int_0^1 x^5\,dx = \frac{\rho}{6}$$

Mass: $\rho \int_0^1 \int_0^{x^3} dy\,dx = \rho \int_0^1 x^3\,dx = \frac{\rho}{4}$

$R_y = \sqrt{\dfrac{\rho}{6}\,\dfrac{4}{\rho}} = \sqrt{\dfrac{2}{3}} = \dfrac{\sqrt{6}}{3}$

13.

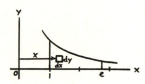

$$I_y = \rho \int_1^e \int_0^{1/x} x^2\,dy\,dx = \rho \int_1^e x^2 y\big|_0^{1/x}\,dx = \rho \int_1^e x\,dx = \frac{\rho}{2} x^2\Big|_1^e$$

$$= \frac{\rho}{2}(e^2 - 1)$$

Mass: $\rho \int_1^e \int_0^{1/x} dy\,dx = \rho \int_1^e \frac{1}{x}\,dx = \rho\,\ln|x|\big\|_1^e$

$$= \rho(\ln e - \ln 1)$$
$$= \rho(1 - 0) = \rho$$

$R_y = \sqrt{\dfrac{\rho(e^2 - 1)}{2}\,\dfrac{1}{\rho}} = \sqrt{\dfrac{e^2 - 1}{2}}$

17.

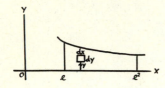

$$y = \frac{1}{\sqrt[3]{x}} = x^{-1/3}$$

Moment of inertia of typical element: $\rho y^2 dy\,dx$

$$I_x = \rho \int_e^{e^2} \int_0^{x^{-1/3}} y^2 dy\,dx = \rho \int_e^{e^2} \frac{1}{3}y^3 \Big|_0^{x^{-1/3}} dx$$

$$= \rho \int_e^{e^2} \frac{1}{3}(x^{-1/3})^3 dx = \frac{\rho}{3} \int_e^{e^2} x^{-1} dx = \frac{\rho}{3} \ln|x| \Big|_e^{e^2}$$

$$= \frac{\rho}{3}(\ln e^2 - \ln e) = \frac{\rho}{3}(2 \ln e - \ln e)$$

$$= \frac{\rho}{3}(2 - 1) = \frac{\rho}{3}$$

21.

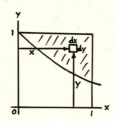

Moment (with respect to y-axis) of typical element: $x\,dy\,dx$

$$M_y = \int_0^1 \int_{e^{-x}}^1 x\,dy\,dx = \int_0^1 xy\Big|_{e^{-x}}^1 dx = \int_0^1 x(1 - e^{-x})\,dx$$

$$= \int_0^1 x\,dx - \int_0^1 xe^{-x} dx$$

$$u = x \qquad dv = e^{-x} dx$$
$$du = dx \qquad v = -e^{-x}$$

$$\frac{1}{2}x^2 \Big|_0^1 - \left[-xe^{-x}\Big|_0^1 + \int_0^1 e^{-x} dx \right] = \frac{1}{2}x^2 \Big|_0^1 - \left[-xe^{-x}\Big|_0^1 - e^{-x}\Big|_0^1 \right]$$

$$= \frac{1}{2}x^2 \Big|_0^1 + xe^{-x}\Big|_0^1 + e^{-x}\Big|_0^1 = \frac{1}{2} + e^{-1} + e^{-1} - 1 = \frac{2}{e} - \frac{1}{2}$$

$$A = \int_0^1 \int_{e^{-x}}^1 dy\,dx = \int_0^1 (1 - e^{-x})\,dx = x + e^{-x}\Big|_0^1 = 1 + e^{-1} - 1 = \frac{1}{e}$$

$$\bar{x} = \frac{2/e - 1/2}{1/e} = 2 - \frac{1}{2}e = \frac{4 - e}{2}$$

$$M_x = \int_0^1 \int_{e^{-x}}^1 y\,dy\,dx = \int_0^1 \frac{1}{2}y^2\Big|_{e^{-x}}^1 dx = \frac{1}{2}\int_0^1 (1 - e^{-2x})\,dx$$

$$u = -2x; \quad du = -2\,dx$$

$$= \frac{1}{2}\Big(x + \frac{1}{2}e^{-2x}\Big)\Big|_0^1 = \frac{1}{2}\Big(1 + \frac{1}{2}e^{-2}\Big) - \frac{1}{2}\Big(\frac{1}{2}\Big)$$

$$= \frac{1}{2} + \frac{1}{4}e^{-2} - \frac{1}{4} = \frac{1}{4e^2} + \frac{1}{4}$$

$$\bar{y} = \frac{1/(4e^2) + 1/4}{1/e} = \frac{1}{4e} + \frac{e}{4} = \frac{e^2 + 1}{4e}$$

25.

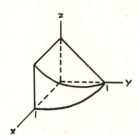 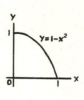

$$\bar{x} = \frac{\displaystyle\int_0^1 \int_0^{1-x^2} \int_0^{1-y} x\,dz\,dy\,dx}{\displaystyle\int_0^1 \int_0^{1-x^2} \int_0^{1-y} dz\,dy\,dx}$$

27.

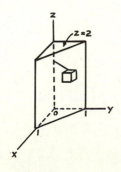

Moment of inertia (with respect to z-axis) of typical element: $\rho(x^2 + y^2)\,dz\,dy\,dx$.

$$I_z = \rho \int_0^1 \int_0^{1-x} \int_0^2 (x^2 + y^2)\,dz\,dy\,dx$$

$$= \rho \int_0^1 \int_0^{1-x} (x^2 + y^2)z\Big|_0^2\,dy\,dx = 2\rho \int_0^1 \int_0^{1-x} (x^2 + y^2)\,dy\,dx$$

$$= 2\rho \int_0^1 \left(x^2 y + \tfrac{1}{3}y^3\right)\Big|_0^{1-x}\,dx = 2\rho \int_0^1 \left[x^2(1 - x) + \tfrac{1}{3}(1 - x)^3\right]dx$$

$$= 2\rho \int_0^1 x^2(1 - x)\,dx + \frac{2\rho}{3} \int_0^1 (1 - x)^3\,dx \qquad\qquad u = 1 - x;\ du = -dx$$

$$= 2\rho\left(\tfrac{1}{3}x^3 - \tfrac{1}{4}x^4\right)\Big|_0^1 - \frac{2\rho}{3}\frac{(1 - x)^4}{4}\Big|_0^1$$

$$= 2\rho\left(\tfrac{1}{3} - \tfrac{1}{4}\right) + \frac{2\rho}{12} = 2\rho\left(\tfrac{1}{12}\right) + \frac{2\rho}{12} = \tfrac{1}{3}\rho$$

29. $$I_y = \rho \int_0^1 \int_{x^2}^x \int_0^{xy} (x^2 + z^2)\,dz\,dy\,dx$$

33. $$V = \int_0^1 \int_0^{1-x} \int_0^2 dz\,dy\,dx = \int_0^1 \int_0^{1-x} z\Big|_0^2\,dy\,dx = 2 \int_0^1 \int_0^{1-x} dy\,dx$$

$$= 2 \int_0^1 y\Big|_0^{1-x}\,dx = 2 \int_0^1 (1 - x)\,dx = 1$$

$$M_{yz} = \int_0^1 \int_0^{1-x} \int_0^2 x\,dz\,dy\,dx = \int_0^1 \int_0^{1-x} xz\Big|_0^2\,dy\,dx$$

$$= 2 \int_0^1 \int_0^{1-x} x\,dy\,dx = 2 \int_0^1 xy\Big|_0^{1-x}\,dx = 2 \int_0^1 x(1 - x)\,dx = \tfrac{1}{3}$$

$$\bar{x} = \frac{M_{yz}}{V} = \tfrac{1}{3}$$

$$M_{xy} = \int_0^1 \int_0^{1-x} \int_0^2 z\,dz\,dy\,dx = \int_0^1 \int_0^{1-x} \tfrac{1}{2}z^2\Big|_0^2\,dy\,dx$$

$$= 2 \int_0^1 \int_0^{1-x} dy\,dx = 2 \int_0^1 (1 - x)\,dx = 1$$

$$\bar{z} = \frac{M_{xy}}{V} = 1$$

5. Trace in xy-plane: $y = 3x^2$

The cylinder $y = 3x^2$ extends along the z-axis (note the missing z-variable).

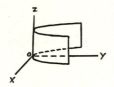

9. $4x^2 + 2y^2 - z^2 = 9$ (hyperboloid of one sheet)

1. Trace in xy-plane ($z = 0$): $4x^2 + 2y^2 = 9$

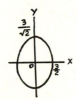

2. Trace in xz-plane ($y = 0$): $4x^2 - z^2 = 9$

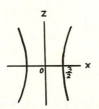

3. Trace in yz-plane $(x = 0)$: $2y^2 - z^2 = 9$

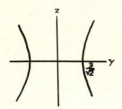

Cross-section: let $z = 4$ in the original equation:

$$4x^2 + 2y^2 - 4^2 = 9 \quad \text{or} \quad 4x^2 + 2y^2 = 25$$

The resulting ellipse is parallel to the xy-plane and 4 units above.

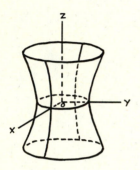

13. $T = 2xy - x^2 - 2y^2 + 3x + 5$

Critical point:

$$\frac{\partial T}{\partial x} = 2y - 2x + 3 = 0$$

$$\underline{\frac{\partial T}{\partial y} = 2x - 4y = 0}$$

$$-2y + 3 = 0 \qquad\qquad \text{adding}$$

$$y = \frac{3}{2}, \quad x = 3$$

$$\frac{\partial^2 T}{\partial x^2} = -2, \quad \frac{\partial^2 T}{\partial y^2} = -4, \quad \frac{\partial^2 T}{\partial x \partial y} = 2$$

$$A = (-2)(-4) - (2)^2 = 4 > 0$$

Since $A > 0$ and $\frac{\partial^2 T}{\partial x^2} < 0$, T has a maximum at $\left(3, \frac{3}{2}\right)$. Finally, substituting in T, we get

$$T\big|_{(3/2,3)} = \frac{19}{2} = 9.5°C \text{ (warmest point)}$$

17.

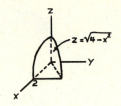

$$V = \int_0^2 \int_0^x \sqrt{4 - x^2} \, dy \, dx = \int_0^2 \sqrt{4 - x^2} \, y\Big|_0^x \, dx = \int_0^2 (4 - x^2)^{1/2} x \, dx$$

$$= -\frac{1}{2} \int_0^2 (4 - x^2)^{1/2} (-2x \, dx) \qquad\qquad u = 4 - x^2; \quad du = -2x \, dx$$

$$= -\frac{1}{2}\left(\frac{2}{3}\right)(4 - x^2)^{3/2}\Big|_0^2 = 0 + \frac{1}{3} \cdot 4^{3/2} = \frac{8}{3}$$

21.

Moment of inertia of typical element: $x^2 \cdot \rho \, dy \, dx$

$$I_y = \rho \int_0^2 \int_0^{4-x^2} x^2 \, dy \, dx = \frac{64\rho}{15}$$

25.

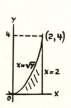 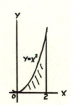

$$\int_0^2 \int_0^{x^2} F(x,y) \, dy \, dx$$

29.

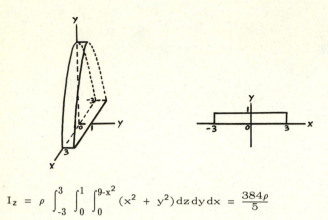

$$I_z = \rho \int_{-3}^{3} \int_{0}^{1} \int_{0}^{9-x^2} (x^2 + y^2)\, dz\, dy\, dx = \frac{384\rho}{5}$$

CHAPTER 10   INFINITE SERIES

Section 10.1

1. $a = 1$, $r = \frac{1}{3}$

$S = \frac{a}{1 - r} = \frac{1}{1 - \frac{1}{3}} = \frac{3}{2}$

5. $a = 1$, $r = \frac{3}{4}$

$S = \frac{1}{1 - \frac{3}{4}} = \frac{1}{\frac{1}{4}} = 4$

9. $\frac{2}{3} - \frac{4}{9} + \frac{8}{27} - \cdots = \frac{2}{3} + \frac{2}{3}\left(-\frac{2}{3}\right) + \frac{2}{3}\left(\frac{4}{9}\right) - \cdots$

$a = \frac{2}{3}$, $r = -\frac{2}{3}$

$S = \frac{2/3}{1 - (-2/3)} = \frac{2/3}{1 + 2/3} = \frac{2/3}{5/3} = \frac{2}{3}\frac{3}{5} = \frac{2}{5}$

13. $0.50707\ldots = 0.5 + 0.007 + 0.00007 + 0.0000007 + \cdots$

$= \frac{1}{2} + \frac{7}{10^3} + \frac{7}{10^5} + \frac{7}{10^7} + \cdots$

$= \frac{1}{2} + \left(\frac{7}{10^3} + \frac{7}{10^3}\frac{1}{10^2} + \frac{7}{10^3}\frac{1}{10^4} + \cdots\right)$

$a = \frac{7}{10^3}$, $r = \frac{1}{10^2}$

$S = \frac{1}{2} + \frac{7/10^3}{1 - 1/10^2} = \frac{1}{2} + \frac{7}{10^3 - 10} = \frac{1}{2} + \frac{7}{990} = \frac{495 + 7}{990}$

$= \frac{502}{990} = \frac{251}{495}$

Section 10.2

1. $\lim_{n \to \infty} \frac{n}{2n + 2} = \lim_{n \to \infty} \frac{1}{2 + 2/n} = \frac{1}{2}$

(n^{th} term does not go to 0)

5. $\displaystyle\int_1^\infty \frac{dx}{(x+1)^2}\,dx = \lim_{b\to\infty} \int_1^b (x+1)^{-2}\,dx$ $\qquad u = x+1;\ du = dx$

$\displaystyle = \lim_{b\to\infty} \left.\frac{(x+1)^{-1}}{-1}\right|_1^b = \lim_{b\to\infty}\left(-\frac{1}{x+1}\right)\Big|_1^b$

$\displaystyle = \lim_{b\to\infty}\left[-\frac{1}{b+1} + \frac{1}{2}\right] = \frac{1}{2}$ $\qquad$ (convergent)

9. $\displaystyle\int_0^\infty \frac{dx}{(2x+2)^2} = \lim_{b\to\infty}\int_0^b (2x+2)^{-2}\,dx$ $\qquad u = 2x+2;\ du = 2dx$

$\displaystyle = \lim_{b\to\infty} \frac{1}{2}\int_0^b (2x+2)^{-2}(2\,dx)$

$\displaystyle = \lim_{b\to\infty}\frac{1}{2}\,\left.\frac{(2x+2)^{-1}}{-1}\right|_0^b = \lim_{b\to\infty}\left(-\frac{1}{2(2x+2)}\right)\Big|_0^b$

$\displaystyle = \lim_{b\to\infty}\left(-\frac{1}{2(2b+2)} + \frac{1}{4}\right) = \frac{1}{4}$ $\qquad$ (convergent)

13. $\displaystyle\int_1^\infty \frac{x}{e^x}\,dx = \int_1^\infty xe^{-x}\,dx$

$\qquad\qquad u = x \qquad dv = e^{-x}dx$

$\qquad\qquad du = dx \qquad v = -e^{-x}$

$\displaystyle \lim_{b\to\infty}\left[-xe^{-x}\big|_1^b + \int_1^b e^{-x}\,dx\right]$

$\displaystyle = \lim_{b\to\infty}\left[-xe^{-x}\big|_1^b - e^{-x}\big|_1^b\right] = \lim_{b\to\infty}\left[\left(-be^{-b} + e^{-1}\right) - \left(e^{-b} - e^{-1}\right)\right]$

$\displaystyle = \lim_{b\to\infty}\left[\left(-\frac{b}{e^b} + \frac{1}{e}\right) - \left(\frac{1}{e^b} - \frac{1}{e}\right)\right] = \frac{2}{e}$ (convergent)

17. $\displaystyle\int_0^\infty \frac{x}{(x^2+2)^2} = \frac{1}{4}$ $\qquad\qquad$ (convergent)

21. $\displaystyle\frac{1}{n^2+1} < \frac{1}{n^2}$ $\qquad$ (converges by comparison to p-series with p = 2)

25. $\displaystyle\frac{1}{n^3+2} < \frac{1}{n^3}$

29. $\displaystyle\frac{1}{3^n+1} < \frac{1}{3^n}$ $\qquad$ (converges by comparison to geometric series)

33. $\displaystyle\frac{1}{\sqrt{n}-1} = \frac{1}{n^{1/2}-1} > \frac{1}{n^{1/2}}$

(diverges by comparison to p-series with p = $\frac{1}{2}$)

37. $\displaystyle\frac{1+\sin n}{n^3} \le \frac{2}{n^3} < \frac{1}{n^2}$ for n > 2

41. $\lim\limits_{n\to\infty} \dfrac{a_{n+1}}{a_n} = \lim\limits_{n\to\infty} \dfrac{1/3^{n+1}}{1/3^n} = \lim\limits_{n\to\infty} \dfrac{1}{3^{n+1}} \dfrac{3^n}{1} = \lim\limits_{n\to\infty} \dfrac{1}{3 \cdot 3^n} \dfrac{3^n}{1}$

$$= \lim\limits_{n\to\infty} \tfrac{1}{3} = \tfrac{1}{3} < 1 \qquad\qquad \text{(convergent)}$$

45. $\lim\limits_{n\to\infty} \dfrac{a_{n+1}}{a_n} = \lim\limits_{n\to\infty} \dfrac{\dfrac{4^{n+1}}{(n+1)!}}{\dfrac{4^n}{n!}} = \lim\limits_{n\to\infty} \dfrac{4^{n+1}}{(n+1)!} \dfrac{n!}{4^n}$

$= \lim\limits_{n\to\infty} \dfrac{4^n \cdot 4}{(n+1)n!} \dfrac{n!}{4^n} = \lim\limits_{n\to\infty} \dfrac{4}{n+1} = 0 < 1 \text{ (convergent)}$

49. $\lim\limits_{n\to\infty} \dfrac{a_{n+1}}{a_n} = \lim\limits_{n\to\infty} \dfrac{(n+1)!/7^{n+1}}{n!/7^n} = \lim\limits_{n\to\infty} \dfrac{(n+1)!}{7^{n+1}} \dfrac{7^n}{n!}$

$$= \lim\limits_{n\to\infty} \dfrac{(n+1)n!}{7 \cdot 7^n} \dfrac{7^n}{n!} = \lim\limits_{n\to\infty} \dfrac{n+1}{7} = \infty \qquad \text{(divergent)}$$

53. $a_{n+1} = \dfrac{1 \cdot 4 \cdot 7 \cdots (3n-2)(3n+1)}{2 \cdot 4 \cdot 6 \cdots (2n)(2n+2)}$

$\dfrac{a_{n+1}}{a_n} = \dfrac{1 \cdot 4 \cdot 7 \cdots (3n-2)(3n+1)}{2 \cdot 4 \cdot 6 \cdots (2n)(2n+2)} \cdot \dfrac{2 \cdot 4 \cdot 6 \cdots (2n)}{1 \cdot 4 \cdot 7 \cdots (3n-2)}$

$$= \dfrac{3n+1}{2n+2} \to \dfrac{3}{2} > 1 \qquad \text{(divergent)}$$

Section 10.3

1. $f(x) = \sin x$ $\qquad\qquad$ $f(0) = 0$

$\quad f'(x) = \cos x$ $\qquad\qquad$ $f'(0) = 1$

$\quad f''(x) = -\sin x$ $\qquad\qquad$ $f''(0) = 0$

$\quad f'''(x) = -\cos x$ $\qquad\qquad$ $f'''(0) = -1$

$\quad f^{(4)}(x) = \sin x$ $\qquad\qquad$ $f^{(4)}(0) = 0$

$\quad f^{(5)}(x) = \cos x$ $\qquad\qquad$ $f^{(5)}(0) = 1$

$\sin x = 0 + 1x + \dfrac{0}{2!}x^2 - \dfrac{1}{3!}x^3 + \dfrac{0}{4!}x^4 + \dfrac{1}{5!}x^5 + \cdots$

$$= x - \dfrac{x^3}{3!} + \dfrac{x^5}{5!} - \cdots$$

7. $f(x) = \ln(1 + x)$ $\qquad\qquad\qquad$ $f(0) = 0$

$\quad f'(x) = \dfrac{1}{x+1} = (x+1)^{-1}$ $\qquad\quad$ $f'(0) = 1$

$\quad f''(x) = -1(x+1)^{-2}$ $\qquad\qquad\quad$ $f''(0) = -1$

$\quad f'''(x) = 1 \cdot 2(x+1)^{-3}$ $\qquad\qquad$ $f'''(0) = 2!$

$$f^{(4)}(x) = -1 \cdot 2 \cdot 3(x + 1)^{-4} \qquad\qquad f^{(4)}(0) = -3!$$
$$f^{(5)}(x) = 1 \cdot 2 \cdot 3 \cdot 4(x - 1)^{-5} \qquad f^{(5)}(0) = 4!$$
$$f^{(6)}(x) = -1 \cdot 2 \cdot 3 \cdot 4 \cdot 5(x - 1)^6 \qquad f^{(6)}(0) = -5!$$

$$\ln(1 + x) = 0 + 1x + \tfrac{-1}{2!}x^2 + \tfrac{2!}{3!}x^3 + \tfrac{-3!}{4!}x^4 + \tfrac{4!}{5!}x^5 + \cdots$$

$$= x - \tfrac{x^2}{2} + \tfrac{x^3}{3} - \tfrac{x^4}{4} + \tfrac{x^5}{5} - \cdots$$

11.
$$f(x) = \text{Arctan } x \qquad\qquad f(0) = 0$$

$$f'(x) = \frac{1}{1 + x^2} = (1 + x^2)^{-1} \qquad f'(0) = 1$$

$$f''(x) = -(1 + x^2)(2x) = -\frac{2x}{(1 + x^2)^2} \qquad f''(0) = 0$$

$$f'''(x) = \frac{6x^2 - 2}{(1 + x^2)^3} \qquad\qquad f'''(0) = -2$$

$$f^{(4)}(x) = \frac{4!(x - x^3)}{(1 + x^2)^4} \qquad\qquad f^{(4)}(0) = 0$$

$$f^{(5)}(x) = \frac{4!(5x^4 - 10x^2 + 1)}{(1 + x^2)^5} \qquad f^{(5)}(0) = 4!$$

$$\text{Arctan } x = 0 + 1x + \tfrac{0}{2!}x^2 + \tfrac{-2}{3!}x^3 + \tfrac{0}{4!}x^4 + \tfrac{4!}{5!}x^5 + \cdots$$

$$= x - \tfrac{x^3}{3} + \tfrac{x^5}{5} - \cdots$$

13. This difficult series is best obtained indirectly by using the binomial theorem to expand the derivative of Arcsin x and integrating the resulting series:

$$\frac{d}{dx} \text{Arcsin } x = \frac{1}{\sqrt{1 - x^2}} = (1 - x^2)^{-1/2}$$

$$= \left[1 + (-x^2)\right]^{-1/2}$$

$$= 1 - \tfrac{1}{2}(-x^2) + \frac{\left(-\tfrac{1}{2}\right)\left(-\tfrac{3}{2}\right)}{2!}(-x^2)^2 + \frac{\left(-\tfrac{1}{2}\right)\left(-\tfrac{3}{2}\right)\left(-\tfrac{5}{2}\right)}{3!}(-x^2)^3$$

$$+ \frac{\left(-\tfrac{1}{2}\right)\left(-\tfrac{3}{2}\right)\left(-\tfrac{5}{2}\right)\left(-\tfrac{7}{2}\right)}{4!}(-x^2)^4$$

$$+ \cdots$$

$$= 1 + \tfrac{1}{2}x^2 + \frac{\tfrac{1}{2} \cdot \tfrac{3}{2}}{2!}x^4 + \frac{\tfrac{1}{2} \cdot \tfrac{3}{2} \cdot \tfrac{5}{2}}{3!}x^6 + \frac{\tfrac{1}{2} \cdot \tfrac{3}{2} \cdot \tfrac{5}{2} \cdot \tfrac{7}{2}}{4!}x^8 + \cdots$$

Integrating term by term yields the desired series:

$$\int_0^x \frac{d}{dx} \text{Arcsin } x dx = \text{Arcsin } x$$

$$= x + \frac{1}{2} \cdot \frac{1}{3}x^3 + \frac{\frac{1}{2} \cdot \frac{3}{2}}{2!} \frac{1}{5}x^5 + \frac{\frac{1}{2} \cdot \frac{3}{2} \cdot \frac{5}{2}}{3!} \frac{1}{7}x^7$$

$$+ \frac{\frac{1}{2} \cdot \frac{3}{2} \cdot \frac{5}{2} \cdot \frac{7}{2}}{4!} \frac{1}{9}x^9 + \ldots$$

$$= x + \frac{1 \cdot x^3}{2 \cdot 3} + \frac{1 \cdot 3 \cdot x^5}{2 \cdot 4 \cdot 5} + \frac{1 \cdot 3 \cdot 5 \cdot x^7}{2 \cdot 4 \cdot 6 \cdot 7}$$

$$+ \frac{1 \cdot 3 \cdot 5 \cdot 7 \cdot x^9}{2 \cdot 4 \cdot 6 \cdot 8 \cdot 9} + \ldots$$

Section 10.4

1. $\sin x = x - \frac{x^3}{3!} + \frac{x^5}{5!} - \ldots$.   Replacing x by 3x,

$$\sin 3x = 3x - \frac{(3x)^3}{3!} + \frac{(3x)^5}{5!} - \ldots = 3x - \frac{3^3x^3}{3!} + \frac{3^5x^5}{5!} - \ldots$$

5. $\cos x = 1 - \frac{x^2}{2!} + \frac{x^4}{4!} - \ldots$.   Replacing x by $\sqrt{x}$,

$$\cos \sqrt{x} = 1 - \frac{(\sqrt{x})^2}{2!} + \frac{(\sqrt{x})^4}{4!} - \ldots = 1 - \frac{x}{2!} + \frac{x^2}{4!} - \ldots$$

9. $\ln(1 + x) = x - \frac{x^2}{2} + \frac{x^3}{3} - \frac{x^4}{4} + \ldots$.   Replacing x by x^2,

$$\ln(1 + x^2) = x^2 - \frac{(x^2)^2}{2} + \frac{(x^2)^3}{3} - \frac{(x^2)^4}{4} + \ldots$$

$$= x^2 - \frac{x^4}{2} + \frac{x^6}{3} - \frac{x^8}{4} + \ldots$$

13. $\frac{d}{dx} \ln(1 + x) = \frac{1}{x + 1} = \frac{1}{1 + x}$

$$= 1 + (-x) + (-x)^2 + (-x)^3 + (-x)^4 + \ldots$$

$$= 1 - x + x^2 - x^3 + x^4 - \ldots$$

(geometric series with r = -x)

$$\ln(1 + x) = \int_0^x \frac{dx}{1 + x} = \int_0^x (1 - x + x^2 - x^3 + x^4 - \ldots) dx$$

$$= x - \frac{x^2}{2} + \frac{x^3}{3} - \frac{x^4}{4} + \frac{x^5}{5} - \ldots$$

17. $\ln(1 + x^2)^3 = 3 \ln(1 + x^2)$

$$= 3\left(x^2 - \frac{x^4}{2} + \frac{x^6}{3} - \frac{x^8}{4} + \ldots\right) \qquad \text{by Exercise 9}$$

21. $-\sqrt{3} + j$: $r = 2$, $\theta = 150° = \frac{5\pi}{6}$. Thus

$-\sqrt{3} + j = 2e^{5\pi j/6}$

25. $-2 + 2j$: $r = \sqrt{8} = 2\sqrt{2}$, $\theta = 135° = \frac{3\pi}{4}$. Thus

$-2 + 2j = 2\sqrt{2}\,e^{3\pi j/4}$

Section 10.5

1. $\sin x = x - \frac{x^3}{3!} + \frac{x^5}{5!} - \frac{x^7}{7!} + \cdots$

$\sin (0.7) = 0.7 - \frac{(0.7)^3}{3!} + \frac{(0.7)^5}{5!}$ (three terms)

$\qquad\qquad = 0.644234$

max. error (fourth term): $-\frac{(0.7)^7}{7!} = -0.000016$

    (a)   0.644234   sum of first three terms
             $\underline{-0.000016}$ error
    (b)   0.644218

The values of (a) and (b) agree to four decimal places: 0.6442.

5. $e^x = 1 + x + \frac{x^2}{2!} + \frac{x^3}{3!} + \frac{x^4}{4!} + \cdots$

$e^{-0.2} = 1 + (-0.2) + \frac{(-0.2)^2}{2!} + \frac{(-0.2)^3}{3!}$ (four terms)

$\qquad\qquad = 0.818667$

max. error (fifth term): $\frac{(-0.2)^4}{4!} = 0.000067$

    (a)   0.818667   sum of first four terms
             $\underline{0.000067}$ error
    (b)   0.818734

The values of (a) and (b) agree to four decimal places (after rounding off):   0.8187.

9. $\ln(1 + x) = x - \frac{x^2}{2} + \frac{x^3}{3} - \frac{x^4}{4} + \cdots$

$\ln 1.1 = \ln(1 + 0.1) = 0.1 - \frac{(0.1)^2}{2} + \frac{(0.1)^3}{3}$ (three terms)

$\qquad\qquad\qquad = 0.095333$

max. error (fourth term): $-\frac{(0.1)^4}{4} = -0.000025$

    (a)   0.095333   sum of first three terms
             $\underline{-0.000025}$ error
    (b)   0.095308

The values of (a) and (b) agree to four decimal places: 0.0953

13. $\dfrac{1 - \cos x}{x} = \left[1 - \left(1 - \dfrac{x^2}{2!} + \dfrac{x^4}{4!} - \dfrac{x^6}{6!} + \dfrac{x^8}{8!} - \dots \right) \right] / x$

$$= \dfrac{x}{2!} - \dfrac{x^3}{4!} + \dfrac{x^5}{6!} - \dfrac{x^7}{8!} + \dots$$

$$\int_0^{1/2} \dfrac{1 - \cos x}{x}\, dx = \int_0^{1/2} \left(\dfrac{x}{2!} - \dfrac{x^3}{4!} + \dfrac{x^5}{6!} - \dfrac{x^7}{8!} + \dots \right) dx$$

$$= \dfrac{x^2}{2 \cdot 2!} - \dfrac{x^4}{4 \cdot 4!} + \dfrac{x^6}{6 \cdot 6!} - \dfrac{x^8}{8 \cdot 8!} + \dots \Big|_0^{1/2}$$

$$= 0.06185 \quad \text{using three terms}$$

error: $- \dfrac{(0.5)^8}{8 \cdot 8!} = -1.2 \times 10^{-8}$ (fourth term)

17. $e^x = 1 + x + \dfrac{x^2}{2!} + \dfrac{x^3}{3!} + \dfrac{x^4}{4!} + \dots$

$$e^{-x^2} = 1 - x^2 + \dfrac{x^4}{2!} - \dfrac{x^6}{3!} + \dfrac{x^8}{4!} - \dots$$

$$\int_0^{0.3} e^{-x^2}\, dx = x - \dfrac{x^3}{3} + \dfrac{x^5}{5 \cdot 2!} - \dfrac{x^7}{7 \cdot 3!} \Big|_0^{0.3} = 0.29124 \quad \text{using four terms}$$

error: $\dfrac{(0.3)^9}{9 \cdot 4!} = 9.1 \times 10^{-8}$ (fifth term)

21. $\quad f(x) = \cos x \qquad\qquad\qquad f\left(\dfrac{\pi}{6}\right) = \dfrac{\sqrt{3}}{2}$

$\quad f'(x) = -\sin x \qquad\qquad\quad f'\left(\dfrac{\pi}{6}\right) = -\dfrac{1}{2}$

$\quad f''(x) = -\cos x \qquad\qquad f''\left(\dfrac{\pi}{6}\right) = -\dfrac{\sqrt{3}}{2}$

$\quad f'''(x) = \sin x \qquad\qquad\quad f'''\left(\dfrac{\pi}{6}\right) = \dfrac{1}{2}$

$\cos x = \dfrac{\sqrt{3}}{2} + \left(-\dfrac{1}{2}\right)\left(x - \dfrac{\pi}{6}\right) + \dfrac{-\sqrt{3}/2}{2!}\left(x - \dfrac{\pi}{6}\right)^2 + \dfrac{1/2}{3!}\left(x - \dfrac{\pi}{6}\right)^3 + \dots$

$31° = 30° + 1° = \dfrac{\pi}{6} + \dfrac{\pi}{180} = x$

Thus $x - \dfrac{\pi}{6} = \left(\dfrac{\pi}{6} + \dfrac{\pi}{180}\right) - \dfrac{\pi}{6} = \dfrac{\pi}{180}$

$\cos 31° = \dfrac{\sqrt{3}}{2} - \dfrac{1}{2}\left(\dfrac{\pi}{180}\right) - \dfrac{\sqrt{3}}{4}\left(\dfrac{\pi}{180}\right)^2 + \dfrac{1}{12}\left(\dfrac{\pi}{180}\right)^3 = 0.85717$

25. Expand $\ln x$ about $c = 1$

$\quad f(x) = \ln x \qquad\qquad\qquad f(1) = 0$

$\quad f'(x) = \dfrac{1}{x} = x^{-1} \qquad\qquad\, f'(1) = 1$

$\quad f''(x) = -1x^{-2} \qquad\qquad\; f''(1) = -1$

$\quad f'''(x) = 1 \cdot 2x^{-3} \qquad\qquad f'''(1) = 2!$

$\quad f^{(4)}(x) = -1 \cdot 2 \cdot 3x^{-3} \qquad f^{(4)}(1) = -3!$

$$\ln x = 0 + 1(x - 1) + \tfrac{-1}{2!}(x - 1)^2 + \tfrac{2!}{3!}(x - 1)^3 + \tfrac{-3!}{4!}(x - 1)^4 + \ldots$$

$$= (x - 1) - \frac{(x - 1)^2}{2} + \frac{(x - 1)^3}{3} - \frac{(x - 1)^4}{4} + \ldots$$

29. (a) $\sin \theta = \theta - \dfrac{\theta^3}{3!} + \dfrac{\theta^5}{5!} - \dfrac{\theta^7}{7!} + \ldots$

If $\theta \approx 0$, then the higher powers of θ become negligible, so that $\sin \theta \approx \theta$.

(b) If $\sin \theta$ is replaced by θ, then

$$\frac{d^2\theta}{dt^2} = - \frac{g}{L}\theta,$$

that is, the acceleration is directly proportional to the displacement and oppositely directed. By Example 5, Section 6.10, θ has the form

$$\theta = a \cos \sqrt{\frac{g}{L}}\, t$$

(c) The period of the cosine function is

$$\frac{2\pi}{\sqrt{\frac{g}{L}}} = 2\pi \sqrt{\frac{L}{g}}$$

31. $e^x = 1 + x + \dfrac{x^2}{2!} + \dfrac{x^3}{3!} + \dfrac{x^4}{4!} + \dfrac{x^5}{5!} + \dfrac{x^6}{6!} + \ldots$

$e^{-x^2/2} = 1 - \dfrac{1}{2}x^2 + \dfrac{1}{4}\dfrac{x^4}{2!} - \dfrac{1}{8}\dfrac{x^6}{3!} + \dfrac{1}{16}\dfrac{x^8}{4!} - \dfrac{1}{32}\dfrac{x^{10}}{5!} + \dfrac{1}{64}\dfrac{x^{12}}{6!} - \ldots$

$$\frac{1}{\sqrt{2\pi}} \int_0^1 e^{-x^2/2} dx = \frac{1}{\sqrt{2\pi}}\left(x - \frac{1}{2}\frac{x^3}{3} + \frac{1}{4}\frac{x^5}{5 \cdot 2!} - \frac{1}{8}\frac{x^7}{7 \cdot 3!}\right.$$

$$\left. + \frac{1}{16}\frac{x^9}{9 \cdot 4!} - \frac{1}{32}\frac{x^{11}}{11 \cdot 5!}\right)\Big|_0^1$$

$$= \frac{1}{\sqrt{2\pi}}\left(1 - \frac{1}{2}\frac{1}{3} + \frac{1}{4}\frac{1}{5 \cdot 2!} - \frac{1}{8}\frac{1}{7 \cdot 3!} + \frac{1}{16}\frac{1}{9 \cdot 4!}\right.$$

$$\left. - \frac{1}{32}\frac{1}{11 \cdot 5!}\right)$$

$$= 0.3413$$

error: $\dfrac{1}{\sqrt{2\pi}}\dfrac{1}{64}\dfrac{1}{13 \cdot 6!} = 6.7 \times 10^{-7}$

1.

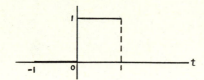

Period: 2p = 2, so that p = 1

$$a_0 = \frac{1}{1} \int_{-1}^{1} f(t)dt = \int_{-1}^{0} 0\,dt + \int_{0}^{1} 1\ dt = 0 + t\Big|_0^1 = 1$$

$$\frac{a_0}{2} = \frac{1}{2}$$

$$a_n = \frac{1}{1} \int_{-1}^{1} f(t)\ \cos\frac{n\pi t}{p}dt = \int_{-1}^{0} 0\,dt + \int_{0}^{1} 1 \cdot \cos\frac{n\pi t}{1}dt$$

$$= \frac{1}{n\pi}\ \sin\ n\pi t\Big|_0^1 = \frac{1}{n\pi}\ \sin\ n\pi = 0 \qquad\qquad u = n\pi t;\ \ du = n\pi dt$$

$$b_n = \frac{1}{1} \int_{-1}^{1} f(t)\ \sin\frac{n\pi t}{p}dt = \int_{-1}^{0} 0\,dt + \int_{0}^{1} 1 \cdot \sin\frac{n\pi t}{1}dt$$

$$= -\frac{1}{n\pi}\ \cos\ n\pi t\Big|_0^1 = -\frac{1}{n\pi}(\cos\ n\pi\ -\ 1) = \frac{1}{n\pi}(1\ -\ \cos\ n\pi)$$

Now recall that

cos 2π = cos 4π = cos 6π = ... = 1

cos π = cos 3π = cos 5π = ... = −1

Hence 1 − cos nπ = 0 whenever n is even and 1 − cos nπ = 1 − (−1) = 2 whenever n is odd. Thus

$$b_n = \begin{cases} 0, & n\ \text{even} \\[2mm] \frac{2}{n\pi}, & n\ \text{odd} \end{cases}$$

It follows that (since p = 1)

$$f(t) = \frac{1}{2} + \frac{2}{\pi}\ \sin\frac{1\pi t}{1} + 0 + \frac{2}{3\pi}\ \sin\frac{3\pi t}{1} + 0 + \frac{2}{5\pi}\ \sin\frac{5\pi t}{1} + \dots$$

$$= \frac{1}{2} + \frac{2}{\pi}\Big(\sin\ \pi t + \frac{1}{3}\ \sin\ 3\pi t + \frac{1}{5}\ \sin\ 5\pi t\ + \dots\Big)$$

5.

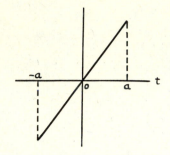

Since the period $2p = 2a$, $p = a$.

$$a_0 = \frac{1}{a} \int_{-a}^{a} t\, dt = \frac{1}{a}\, \frac{1}{2} t^2 \Big|_{-a}^{a} = 0$$

$$a_n = \frac{1}{a} \int_{-a}^{a} t\, \cos \frac{n\pi t}{a}\, dt$$

$$u = t \qquad dv = \cos \frac{n\pi t}{a}\, dt$$

$$du = dt \qquad v = \frac{a}{n\pi} \sin \frac{n\pi t}{a}$$

$$a_n = \frac{1}{a}\left[t\left(\frac{a}{n\pi}\right) \sin \frac{n\pi t}{a}\Big|_{-a}^{a} - \frac{a}{n\pi} \int_{-a}^{a} \sin \frac{n\pi t}{a}\, dt \right]$$

$$= \frac{1}{a}\left[0 - \frac{a}{n\pi}\left(-\frac{a}{n\pi}\right) \cos \frac{n\pi t}{a}\Big|_{-a}^{a} \right] = \frac{a}{n^2\pi^2}\left[\cos \frac{n\pi a}{a} - \cos\left(\frac{-n\pi a}{a}\right) \right]$$

$$= \frac{a}{n^2\pi^2}(\cos n\pi - \cos n\pi) = 0, \text{ since } \cos(-\theta) = \cos \theta.$$

$$b_n = \frac{1}{a} \int_{-a}^{a} t\, \sin \frac{n\pi t}{a}\, dt$$

$$u = t \qquad dv = \sin \frac{n\pi t}{a}\, dt$$

$$du = dt \qquad v = -\frac{a}{n\pi} \cos \frac{n\pi t}{a}$$

$$= \frac{1}{a}\left[-\frac{at}{n\pi} \cos \frac{n\pi t}{a}\Big|_{-a}^{a} + \frac{a}{n\pi} \int_{-a}^{a} \cos \frac{n\pi t}{a}\, dt \right]$$

$$= \frac{1}{a}\left[-\frac{at}{n\pi} \cos \frac{n\pi t}{a}\Big|_{-a}^{a} + \frac{a^2}{n^2\pi^2} \sin \frac{n\pi t}{a}\Big|_{-a}^{a} \right]$$

$$= \frac{1}{a}\left[-\frac{a^2}{n\pi} \cos \frac{n\pi a}{a} - \frac{a^2}{n\pi} \cos\left(\frac{-n\pi a}{a}\right) + 0 \right]$$

$$= -\frac{a}{n\pi}(\cos n\pi + \cos n\pi) = -\frac{2a}{n\pi} \cos n\pi$$

since $\cos(-\theta) = \cos \theta$. Thus

$$b_n = \begin{cases} -\dfrac{2a}{n\pi}, & n \text{ even} \\[2mm] \dfrac{2a}{n\pi}, & n \text{ odd} \end{cases}$$

$$f(t) = \frac{2a}{1\pi} \sin \frac{1\pi t}{a} - \frac{2a}{2\pi} \sin \frac{2\pi t}{a} + \frac{2a}{3\pi} \sin \frac{3\pi t}{a} - \ldots$$

$$= \frac{2a}{\pi}\left(\sin \frac{\pi t}{a} - \frac{1}{2} \sin \frac{2\pi t}{a} + \frac{1}{3} \sin \frac{3\pi t}{a} - \ldots \right)$$

9. Since one period is $[-\pi, \pi]$, we have $p = \pi$. Thus

$$a_0 = \frac{1}{\pi} \int_{-\pi}^{\pi} f(t)\,dt = \frac{1}{\pi} \int_{-\pi}^{0} 0 \cdot dt + \frac{1}{\pi} \int_{0}^{\pi} \sin t \, dt$$

$$= \frac{1}{\pi}(-\cos t)\Big|_0^{\pi} = \frac{1}{\pi}(1 + 1) = \frac{2}{\pi}$$

and

$$\frac{a_0}{2} = \frac{1}{\pi}$$

Next,

$$a_n = \frac{1}{\pi} \int_{-\pi}^{\pi} f(t) \cos \frac{n\pi t}{\pi}\,dt = \frac{1}{\pi} \int_{0}^{\pi} \sin t \cos nt\,dt$$

Now, by formula 63, Table 5, with $m = 1$, we get

$$a_n = \frac{1}{\pi}\left[-\frac{\cos(1+n)t}{2(1+n)} - \frac{\cos(1-n)t}{2(1-n)} \right]_0^{\pi}$$

$$= \frac{1}{2\pi}\left[-\frac{\cos(1+n)\pi}{1+n} - \frac{\cos(1-n)\pi}{1-n} + \frac{1}{1+n} + \frac{1}{1-n} \right] \quad (n \neq 1)$$

Observe that this expression is valid for all n except $n = 1$. So if $n \neq 1$, note that

$$\cos(1+n)\pi = \cos(1-n)\pi = \begin{cases} 1, & \text{for } n \text{ odd} \\[2mm] -1, & \text{for } n \text{ even} \end{cases}$$

For n <u>odd</u>, we have

$$\frac{1}{2\pi}\left[-\frac{1}{1+n} - \frac{1}{1-n} + \frac{1}{1+n} + \frac{1}{1-n} \right] = 0$$

For n <u>even</u>, we have

$$\frac{1}{2\pi}\left[\frac{1}{1+n} + \frac{1}{1-n} + \frac{1}{1+n} + \frac{1}{1-n} \right] = \frac{1}{2\pi}\left[\frac{2}{1+n} + \frac{2}{1-n} \right]$$

$$= \frac{1}{\pi}\left[\frac{1}{1+n} + \frac{1}{1-n} \right] = \frac{1}{\pi}\frac{1-n+1+n}{(1+n)(1-n)} = \frac{1}{\pi}\frac{2}{1-n^2} = -\frac{2}{\pi(n^2-1)}$$

Since these forms are not valid for $n = 1$, the coefficient a_1 must be evaluated separately:

$$a_1 = \frac{1}{\pi} \int_0^{\pi} \sin t \cos t \, dt = \frac{\sin^2 t}{2\pi}\Big|_0^{\pi} = 0 \qquad u = \sin t; \; du = \cos t\,dt$$

Thus
$$a_n = \begin{cases} - \dfrac{2}{\pi(n^2 - 1)}, & n \text{ even} \\ 0, & n \text{ odd} \end{cases}$$

Next,
$$b_n = \tfrac{1}{\pi} \int_{-\pi}^{\pi} f(t) \sin \tfrac{n\pi t}{\pi} dt = \tfrac{1}{\pi} \int_0^{\pi} \sin t \sin nt\, dt$$

By formula 61, Table 5, with m = 1, we get
$$b_n = \tfrac{1}{\pi} \left[- \frac{\sin (1 + n)t}{2(1 + n)} + \frac{\sin (1 - n)t}{2(1 - n)} \right]_0^{\pi} \qquad (n \neq 1)$$

$$= 0, \qquad \text{provided that } n \neq 1$$

If n = 1, we have
$$b_1 = \tfrac{1}{\pi} \int_0^{\pi} \sin t \sin t\, dt = \tfrac{1}{\pi} \int_0^{\pi} \sin^2 t\, dt$$

$$= \tfrac{1}{2\pi} \int_0^{\pi} (1 - \cos 2t)\, dt = \tfrac{1}{2\pi} (t - \tfrac{1}{2} \sin 2t) \Big|_0^{\pi} = \tfrac{1}{2}$$

Substituting the first few values in series (10.33), we obtain

$$f(t) = \tfrac{1}{\pi} + \tfrac{1}{2} \sin \tfrac{1\pi t}{\pi} \qquad\qquad \tfrac{a_0}{2} = \tfrac{1}{\pi}, \ b_1 = \tfrac{1}{2}$$

$$+ \ 0 - \frac{2}{\pi \cdot 3} \cos \tfrac{2\pi t}{\pi} + 0 - \frac{2}{\pi \cdot 15} \cos \tfrac{4\pi t}{\pi}$$

$$+ \ 0 - \frac{2}{\pi \cdot 35} \cos \tfrac{6\pi t}{\pi} + 0 - \frac{2}{\pi \cdot 63} \cos \tfrac{8\pi t}{\pi} + \ldots$$

or

$$f(t) = \tfrac{1}{\pi} + \tfrac{1}{2} \sin t$$

$$- \ \tfrac{2}{\pi} \left(\tfrac{1}{3} \cos 2t + \tfrac{1}{15} \cos 4t + \tfrac{1}{35} \cos 6t + \tfrac{1}{63} \cos 8t + \ldots \right)$$

REVIEW EXERCISES FOR CHAPTER 10

1. $r = -\tfrac{1}{3}, \ a = 1$

   $S = \dfrac{1}{1 + 1/3} = \dfrac{3}{4}$

3. $\lim\limits_{n \to \infty} \dfrac{2n}{4n + 3} = \dfrac{1}{2}$.

   Since the nth term does not approach 0, the series diverges.

5. Suppose we try the ratio test:

$$\lim_{n \to \infty} \frac{a_{n+1}}{a_n} =$$

$$\lim_{n \to \infty} \frac{1}{(n + 1)\ln^2(n + 1)} \frac{n \ln^2 n}{1} = \lim_{n \to \infty} \frac{n}{n + 1} \lim_{n \to \infty} \frac{\ln^2 n}{\ln^2(n + 1)} =$$

$$1 \cdot \lim_{n \to \infty} \frac{\ln^2 n}{\ln^2(n + 1)}. \quad \text{By L'Hospital's rule}$$

$$\lim_{n \to \infty} \frac{\ln^2 n}{\ln^2(n + 1)} = \lim_{n \to \infty} \frac{(2 \ln n)\frac{1}{n}}{[2 \ln(n + 1)]\frac{1}{n + 1}}$$

$$= \lim_{n \to \infty} \frac{n + 1}{n} \lim_{n \to \infty} \frac{\ln n}{\ln(n + 1)} = 1 \cdot \lim_{n \to \infty} \frac{1/n}{1/(n + 1)}$$

$$= \lim_{n \to \infty} \frac{n + 1}{n} = 1 \quad \text{(test fails)}$$

By the integral test

$$\int_2^\infty \frac{dx}{x \ln^2 x} = \lim_{b \to \infty} \int_2^b (\ln x)^{-2} \frac{dx}{x} \qquad\qquad u = \ln x; \ du = \frac{1}{x}dx$$

$$= \lim_{b \to \infty} \frac{(\ln x)^{-1}}{-1}\Big|_2^b = \lim_{b \to \infty} \left(-\frac{1}{\ln x}\right)\Big|_2^b$$

$$= \lim_{b \to \infty} \left(-\frac{1}{\ln b} + \frac{1}{\ln 2}\right) = 0 + \frac{1}{\ln 2} \qquad \text{(series converges)}$$

9. $$\lim_{n \to \infty} \frac{a_{n+1}}{a_n} = \lim_{n \to \infty} \frac{6^{n+1}}{(n + 1)!} \frac{n!}{6^n} = \lim_{n \to \infty} \frac{6 \cdot 6^n}{(n + 1)n!} \frac{n!}{6^n}$$

$$= \lim_{n \to \infty} \frac{6}{n + 1} = 0 < 1$$

The series is therefore convergent by the ratio test.

13. $\sin x = x - \frac{x^3}{3!} + \frac{x^5}{5!} - \ldots$. Replacing x by x^2,

$$\sin x^2 = x^2 - \frac{(x^2)^3}{3!} + \frac{(x^2)^5}{5!} - \ldots = x^2 - \frac{x^6}{3!} + \frac{x^{10}}{5!} - \ldots$$

17. $\cos x = 1 - \frac{x^2}{2!} + \frac{x^4}{4!} - \frac{x^6}{6!} + \ldots$

$$\cos(0.5) = 1 - \frac{(0.5)^2}{2!} + \frac{(0.5)^4}{4!} \qquad \text{(three terms)}$$

$$= 0.877604$$

max. error (fourth term): $- \frac{(0.5)^6}{6!} = -0.000022$

(a)   0.877604   sum of first three terms
     -0.000022   error
(b)   0.877582

The values of (a) and (b) agree to four decimal places (after rounding off): 0.8776

21. $e^x = 1 + x + \frac{x^2}{2!} + \frac{x^3}{3!} + \cdots$

$e^{-0.20t^2} = 1 - 0.20t^2 + \frac{(-0.20t^2)^2}{2!} + \frac{(-0.20t^2)^3}{3!} + \cdots$

$\qquad = 1 - 0.20t^2 + \frac{(0.20)^2}{2!}t^4 - \frac{(0.20)^3}{3!}t^6 + \cdots$

$\int_0^{0.10} \left(1 - 0.20t^2 + \frac{(0.20)^2}{2!}t^4 - \frac{(0.20)^3}{3!}t^6 + \cdots\right)dt$

$= t - 0.20\,\frac{t^3}{3} + \frac{(0.20)^2}{2!}\,\frac{t^5}{5} - \frac{(0.20)^3}{3!}\,\frac{t^7}{7}\Big|_0^{0.10}$

$= 0.10 - 0.20\,\frac{(0.10)^3}{3} + \frac{(0.20)^2}{2!}\,\frac{(0.10)^5}{5}$ (three terms)

$= 0.0999 \approx 0.10$ coulombs (two significant digits)

error: $-\frac{(0.20)^3}{3!}\,\frac{(0.10)^7}{7} = -1.9 \times 10^{-11}$

25. $\lim\limits_{x \to 0} \frac{1 - \cos x}{x} = \lim\limits_{x \to 0} \frac{1 - \left(1 - \frac{x^2}{2!} + \frac{x^4}{4!} - \frac{x^6}{6!} + \cdots\right)}{x}$

$= \lim\limits_{x \to 0} \left(\frac{x}{2!} - \frac{x^3}{4!} + \frac{x^5}{6!} - \cdots\right) = 0$

CHAPTER 11   FIRST-ORDER DIFFERENTIAL EQUATIONS

<u>Section 11.1</u>

1. $y = 2e^{3x}$; $\frac{dy}{dx} = 6e^{3x}$.   Substituting in $\frac{dy}{dx} - 3y = 0$, we get

$6e^{3x} - 3(2e^{3x}) = 0$

so that the solution checks.

5. $y = 2 \cos 2x + 3 \sin 2x$

$y' = -4 \sin 2x + 6 \cos 2x$

$y'' = -8 \cos 2x - 12 \sin 2x$

Substituting in the left side of $y'' + 4y = 0$, we get

$(-8 \cos 2x - 12 \sin 2x) + 4(2 \cos 2x + 3 \sin 2x)$

$\quad = -8 \cos 2x - 12 \sin 2x + 8 \cos 2x + 12 \sin 2x = 0$

(The solution checks.)

9. $y = x^2 - 4$, $\frac{dy}{dx} = 2x$

Substituting in the left side of the given equation, we get

$\quad x \frac{dy}{dx} - y = x(2x) - (x^2 - 4) = 2x^2 - x^2 + 4 = x^2 + 4,$

which is equal to the right side.

13. $y = c \cos 3x - x \cos 3x$.   By the product rule,

$\quad y' = -3c \sin 3x - x(-3 \sin 3x) - \cos 3x \cdot 1$

$\quad\quad = -3c \sin 3x + 3x \sin 3x - \cos 3x$

$\quad y'' = -9c \cos 3x + 3x(3 \cos 3x) + 3 \sin 3x + 3 \sin 3x$

$\quad\quad = -9c \cos 3x + 9x \cos 3x + 6 \sin 3x$

Substituting in the left side of the given equation, we get

$y'' + 9y = (-9c \cos 3x + 9x \cos 3x + 6 \sin 3x)$

$\quad\quad\quad\quad\quad\quad + 9(c \cos 3x - x \cos 3x)$

$\quad\quad\quad = 6 \sin 3x,$ the right side

17. $\frac{dy}{dx} = 3x^2$

$\quad y = x^3 + c$ \quad\quad\quad\quad integrating

now let $x = 2$ and $y = 5$:

$$5 = 2^3 + c \quad\quad \text{or} \quad\quad c = -3$$

So

$$y = x^3 - 3$$

21. $\dfrac{d^2y}{dx^2} = e^x$

$\dfrac{dy}{dx} = e^x + c_1$ integrating

$y = e^x + c_1 x + c_2$ integrating again

Substituting $(0,0)$ and $(1,1)$, respectively, we obtain the system of equations

$$0 = 1 + c_2$$
$$1 = e + c_1 + c_2$$

Thus $c_2 = -1$. From the second equation, we get

$$1 = e + c_1 - 1 \quad\quad \text{or} \quad\quad c_1 = 2 - e$$

The solution is therefore given by

$$y = e^x + (2 - e)x - 1$$

Section 11.2

1. $x^2 dx + y dy = 0$

$\displaystyle\int x^2 dx + \int y dy = c_1$

$\dfrac{x^3}{3} + \dfrac{y^2}{2} = c_1$

$2x^3 + 3y^2 = 6c_1$ multiplying by 6

$2x^3 + 3y^2 = c$ (let $c = 6c_1$)

5. $2x \, dx + (1 + x^2) dy = 0$

$\dfrac{2x \, dx}{1 + x^2} + dy = 0$ dividing by $1 + x^2$

$\displaystyle\int \dfrac{2x \, dx}{1 + x^2} + \int dy = c$

$\displaystyle\int \dfrac{du}{u} + \int dy = c$ $u = 1 + x^2$; $du = 2x dx$

$\ln(1 + x^2) + y = c$

9. $2y \, dx + 3x \, dy = 0$

$\dfrac{2 \, dx}{x} + \dfrac{3 \, dy}{y} = 0$ dividing by xy

$\displaystyle\int \dfrac{2 \, dx}{x} + \int \dfrac{3 \, dy}{y} = c_2$ Form: $\displaystyle\int \dfrac{du}{u}$

220

$$2 \ln|x| + 3 \ln|y| = c_2$$
$$\ln x^2 + \ln|y|^3 = \ln c_1 \qquad \text{(let } \ln c_1 = c_2, \ c_1 > 0\text{)}$$
$$\ln|x^2 y^3| = \ln c_1 \qquad \ln A + \ln B = \ln AB$$
$$|x^2 y^3| = c_1, \ c_1 > 0$$
$$x^2 y^3 = \pm c_1$$
$$x^2 y^3 = c, \ c \neq 0 \qquad \text{(let } c = \pm c_1\text{)}$$

13. $$dx - y\,dx + x\,dy = 0$$
$$(1 - y)dx + x\,dy = 0 \qquad \text{factoring } dx$$
$$\frac{dx}{x} + \frac{dy}{1 - y} = 0 \qquad \text{dividing by } x(1 - y)$$

$$\int \frac{dx}{x} + \int \frac{dy}{1 - y} = c_1$$

$$\int \frac{dx}{x} - \int \frac{-dy}{1 - y} = c_1 \qquad u = 1 - y; \ du = -dy$$

$$\ln|x| - \ln|1 - y| = \ln c_2, \ c_2 > 0 \qquad \text{letting } \ln c_2 = c_1$$

$$\ln\left|\frac{x}{1 - y}\right| = \ln c_2 \qquad \ln A - \ln B = \ln \frac{A}{B}$$

$$\left|\frac{x}{1 - y}\right| = c_2$$

$$\frac{x}{1 - y} = \pm c_2$$

$$\frac{x}{1 - y} = c, \ c \neq 0 \qquad \text{letting } c = \pm c_2$$

$$x = c(1 - y)$$

17. $$dx + (2\cos^2 x - y\cos^2 x)dy = 0$$
$$dx + \cos^2 x(2 - y)dy = 0 \qquad \text{factoring } \cos^2 x$$

$$\frac{dx}{\cos^2 x} + (2 - y)dy = 0 \qquad \text{dividing by } \cos^2 x$$

$$\int \frac{dx}{\cos^2 x} + \int (2 - y)dy = c_1$$

$$\int \sec^2 x\,dx + \int (2 - y)dy = c_1 \qquad \frac{1}{\cos x} = \sec x$$

$$\tan x + 2y - \tfrac{1}{2}y^2 = c_1$$

$$2\tan x + 4y - y^2 = 2c_1 \qquad \text{multiplying by } 2$$

$$2\tan x + 4y - y^2 = c \qquad \text{letting } c = 2c_1$$

21. $$\sqrt{v^2 + 1}\,dt + vt^2\,dv = 0$$

$$\frac{dt}{t^2} + \frac{v\,dv}{\sqrt{v^2 + 1}} = 0 \qquad \text{dividing by } t^2\sqrt{v^2 + 1}$$

$$\int \frac{dt}{t^2} + \int \frac{v \, dv}{(v^2 + 1)^{1/2}} = c$$

$$\int t^{-2} dt + \int (v^2 + 1)^{-1/2} v \, dv = c \qquad\qquad u = v^2 + 1; \ du = 2v \, dv$$

$$\int t^{-2} dt + \frac{1}{2} \int (v^2 + 1)^{-1/2} (2v \, dv) = c$$

$$\frac{t^{-1}}{-1} + \frac{1}{2} \frac{(v^2 + 1)^{1/2}}{1/2} = c$$

$$-\frac{1}{t} + \sqrt{v^2 + 1} = c$$

$$-1 + t\sqrt{v^2 + 1} = ct \qquad\qquad \text{multiplying by t}$$

25. $(y^2 - 1)\cos x \, dx + 2y \sin x \, dy = 0$

$$\frac{\cos x}{\sin x} dx + \frac{2y}{y^2 - 1} dy = 0$$

$\ln|\sin x| + \ln|y^2 - 1| = \ln c_1, \ c_1 > 0 \qquad \text{integrating}$

$\quad \ln|(\sin x)(y^2 - 1)| = \ln c_1 \qquad\qquad \ln A + \ln B = \ln AB$

$\qquad |(y^2 - 1)\sin x| = c_1, \ c_1 > 0$

$\qquad (y^2 - 1)\sin x = \pm c_1$

$\qquad (y^2 - 1)\sin x = c, \ c \neq 0$

29. $(e^x \tan y + \tan y)\dfrac{dy}{dx} + e^x = 0$

$\quad (e^x + 1)\tan y \, dy + e^x dx = 0$

$$\int \tan y \, dy + \int \frac{e^x}{e^x + 1} dx = \ln c_1 \qquad\qquad u = e^x + 1; \ du = e^x dx$$

$\ln|\sec y| + \ln(e^x + 1) = \ln c_1, \ c_1 > 0 \qquad \text{integrating}$

$\quad \ln|(e^x + 1)\sec y| = \ln c_1 \qquad\qquad \ln A + \ln B = \ln AB$

$\qquad |(e^x + 1)\sec y| = c_1, \ c_1 > 0$

$\qquad (e^x + 1)\sec y = \pm c_1$

$\qquad (e^x + 1)\sec y = c, \ c \neq 0$

33. $x \, dy - y \, dx = 0, \ y = 2 \text{ when } x = 1$

$\dfrac{dy}{y} - \dfrac{dx}{x} = 0$

$\ln y - \ln x = \ln c \qquad\qquad\qquad \text{integrating}$

$$\ln\left(\frac{y}{x}\right) = \ln c$$

$$\frac{y}{x} = c$$

Thus $y = cx$. Substituting the given values, we get

$2 = c \cdot 1 \qquad \text{or} \qquad y = 2x$

37. $dx + x \tan y \, dy = 0$, $y = 0$ when $x = 1$

$\dfrac{dx}{x} + \tan y \, dy = 0$

$\ln x + \ln \sec y = \ln c$ integrating

$\quad \ln x \sec y = \ln c$ $\ln A + \ln B = \ln AB$

$\quad\quad x \sec y = c$

If $x = 1$ and $y = 0$, we get $1 \cdot \sec 0 = 1 = c$, so that

$x \sec y = 1$

$\quad x = \dfrac{1}{\sec y}$

$\quad x = \cos y$

Section 11.3

1. $\dfrac{dy}{dx} + 1y = 1$; I.F. $= e^{\int 1\,dx} = e^x$

$e^x\left(\dfrac{dy}{dx} + y\right) = e^x \cdot 1$ multiplying by e^x

$\dfrac{d}{dx}(ye^x) = e^x$ by (11.9)

$\quad ye^x = \displaystyle\int e^x dx = e^x + c$

$\quad\quad y = 1 + ce^{-x}$

5. $xdy + (y - x)dx = 0$

$x\dfrac{dy}{dx} + y - x = 0$

$\dfrac{dy}{dx} + \left(\dfrac{1}{x}\right)y = 1$ I.F. $= e^{\int (1/x)dx} = e^{\ln x} = x$

$x\left(\dfrac{dy}{dx} + \dfrac{y}{x}\right) = x$ multiplying by I.F.

$\dfrac{d}{dx}(xy) = x$ derivative of the product of y and I.F.

$\quad xy = \tfrac{1}{2}x^2 + c$ integrating

$\quad\quad y = \tfrac{1}{2}x + \dfrac{c}{x}$

9. $\dfrac{dy}{dx} - \dfrac{2}{x}y = x^2 \sec^2 x$

I.F. $= e^{\int (-2/x)dx} = e^{-2\ln x} = e^{\ln x^{-2}} = x^{-2} = \dfrac{1}{x^2}$

$\dfrac{1}{x^2}\left(\dfrac{dy}{dx} - \dfrac{2}{x}y\right) = \dfrac{1}{x^2}(x^2 \sec^2 x)$ multiplying by $\dfrac{1}{x^2}$

$\dfrac{d}{dx}\left(\dfrac{1}{x^2}y\right) = \sec^2 x$ derivative of the product of y and I.F.

223

$$\frac{1}{x^2}\, y = \tan x + c \qquad\qquad \text{integrating}$$

$$y = x^2 \tan x + cx^2$$

13. $xy' - 2y = x^3 e^x$

$$\frac{dy}{dx} - \frac{2}{x} y = x^2 e^x \qquad\qquad \text{I.F.} = e^{\int (-2/x)dx} = e^{-2\ln x} = e^{\ln x^{-2}} = x^{-2} = \frac{1}{x^2}$$

$$\frac{1}{x^2}\left(\frac{dy}{dx} - \frac{2}{x}y\right) = \frac{1}{x^2}(x^2 e^x) \qquad\qquad \text{multiplying by I.F.}$$

$$\frac{d}{dx}\left(\frac{1}{x^2}y\right) = e^x$$

$$\frac{1}{x^2}y = e^x + c \qquad\qquad \text{integrating}$$

$$y = x^2 e^x + cx^2$$

17. $y' - y \tan x - \cos x = 0$

$$\frac{dy}{dx} - (\tan x)y = \cos x \qquad\qquad \text{I.F.} = e^{-\int \tan x\, dx} = e^{\ln \cos x} = \cos x$$

$$\cos x\left(\frac{dy}{dx} - y \tan x\right) = \cos^2 x \qquad\qquad \text{multiplying by I.F.}$$

$$\frac{d}{dx}(y \cos x) = \frac{1}{2}(1 + \cos 2x) \qquad \text{half-angle formula}$$

$$y \cos x = \frac{1}{2} \int (1 + \cos 2x)dx \qquad u = 2x;\ du = 2dx$$

$$y \cos x = \frac{1}{2}x + \frac{1}{4} \sin 2x + c$$

$$4y \cos x = 2x + \sin 2x + c$$

21. $y' - \frac{1}{x}y = x^2 \sin x^2 \qquad \text{I.F.} = e^{\int (-1/x)dx} = e^{-\ln x} = e^{\ln x^{-1}} = x^{-1} = \frac{1}{x}$

$$\frac{1}{x}\left(\frac{dy}{dx} - \frac{1}{x}y\right) = x \sin x^2 \qquad\qquad \text{multiplying by } \frac{1}{x}$$

$$\frac{d}{dx}\left(\frac{y}{x}\right) = x \sin x^2$$

$$\frac{y}{x} = \int x \sin x^2 dx = \frac{1}{2} \int \sin x^2 (2x)dx \qquad u = x^2;\ du = 2xdx$$

$$\frac{y}{x} = -\frac{1}{2} \cos x^2 + c$$

$$y = -\frac{1}{2}x \cos x^2 + cx$$

25. $t\dfrac{dr}{dt} + r = t \ln t$

$$\frac{dr}{dt} + \frac{1}{t}r = \ln t \qquad\qquad \text{dividing by } t$$

224

$$\text{I.F.} = e^{\int (1/t)dt} = e^{\ln t} = t$$

$$t\left(\frac{dr}{dt} + \frac{1}{t}r\right) = t \ln t \qquad \text{multiplying by I.F.}$$

$$\frac{d}{dt}(tr) = t \ln t$$

$$u = \ln t \qquad dv = t dt$$

$$du = \frac{1}{t}dt \qquad v = \frac{1}{2}t^2$$

$$tr = \frac{1}{2}t^2 \ln t - \int \frac{1}{t} \cdot \frac{1}{2}t^2 dt$$

$$tr = \frac{1}{2}t^2 \ln t - \int \frac{1}{2}t dt$$

$$tr = \frac{1}{2}t^2 \ln t - \frac{1}{4}t^2 + c$$

$$r = \frac{1}{2}t \ln t - \frac{1}{4}t + \frac{c}{t} \qquad \text{dividing by } t$$

29. $\frac{dy}{dx} + y = 6e^{-x}$, $y = 2$ when $x = 0$

$$\text{I.F.} = e^x$$

$$e^x\left(\frac{dy}{dx} + y\right) = e^x\left(6e^{-x}\right) \qquad \text{multiplying by I.F.}$$

$$\frac{d}{dx}\left(ye^x\right) = 6$$

$$ye^x = 6x + c \qquad \text{integrating}$$

$$y = 6xe^{-x} + ce^{-x} \qquad \text{multiplying by } e^{-x}$$

$$2 = 0 + c \cdot 1 \quad \text{or} \quad c = 2 \qquad y = 2, \ x = 0$$

$$y = 6xe^{-x} + 2e^{-x}$$

$$y = 2e^{-x}(3x + 1)$$

Section 11.4

1. $L \frac{di}{dt} + Ri = e(t)$

$$0.2 \frac{di}{dt} + 5i = 5$$

$$\frac{di}{dt} + \frac{5}{0.2}i = \frac{5}{0.2}$$

$$\frac{di}{dt} + 25i = 25 \qquad \text{I.F.} = e^{25t}$$

$$e^{25t}\left(\frac{di}{dt} + 25i\right) = 25e^{25t}$$

$$\frac{d}{dt}(ie^{25t}) = 25e^{25t}$$

$$ie^{25t} = \int 25 e^{25t} dt \qquad u = 25t; \ du = 25 dt$$

225

$$ie^{25t} = e^{25t} + c$$

$$i = 1 + ce^{-25t}$$

Initial condition: if $t = 0$, $i = 0$. Thus $0 = 1 + c$ or $c = -1$.
$i = 1 - e^{-25t}$

5. By Example 1, $N = N_0 e^{kt}$.

Since the half-life is 100 hours, we have $N = \frac{N_0}{2}$ when $t = 100$:

$$\tfrac{1}{2} N_0 = N_0 e^{100k}$$

$$\tfrac{1}{2} = e^{100k}$$

$$\ln \tfrac{1}{2} = \ln e^{100k} = 100k, \text{ or } k = \frac{\ln(1/2)}{100} = -0.00693$$

Solution: $N = N_0 e^{-0.00693t}$

Now we need to find t such that $N = 0.90N_0$:

$$0.90N_0 = N_0 e^{-0.00693t}$$

$$0.90 = e^{-0.00693t}$$

$$\ln(0.90) = -0.00693t$$

$$t = \frac{\ln(0.90)}{-0.00693} = 15.2 \text{ h}$$

9. By Example 1, $N = N_0 e^{kt}$.

Since the half-life is 5600 years, we have $N = \tfrac{1}{2}N_0$ when
$t = 5600$ years:

$$\tfrac{1}{2}N_0 = N_0 e^{5600k}$$

$$\tfrac{1}{2} = e^{5600k}$$

$$\ln \tfrac{1}{2} = \ln e^{5600k} = 5600k$$

$$k = \frac{\ln(1/2)}{5600} = -0.000124$$

Solution: $N = N_0 e^{-0.000124t}$
Now find t such that $N = 0.25N_0$:

$$0.25N_0 = N_0 e^{-0.000124t}$$

$$0.25 = e^{-0.000124t} \qquad\qquad \text{dividing by } N_0$$

$$\ln(0.25) = -0.000124t$$

$$t = \frac{\ln(0.25)}{-0.000124} = 11,200 \text{ years}$$

13. Since $M_T = 70°$, the equation is

$$\frac{dT}{dt} = -k(T - 70)$$

$$\frac{dT}{dt} + kT = 70k \qquad I.F. = e^{kt}$$

$$\frac{d}{dt}(Te^{kt}) = 70ke^{kt}$$

$$Te^{kt} = 70 \int e^{kt}(kdt) \qquad\qquad u = kt; \; du = kdt$$

$$Te^{kt} = 70e^{kt} + c$$

$$T = 70 + ce^{-kt}$$

Initial condition: if $t = 0$, $T = 20$:

$20 = 70 + c \quad$ or $\quad c = -50$

and

$T = 70 - 50e^{-kt}$

From the second condition, $T = 35$ when $t = 2$, we have

$35 = 70 - 50e^{-2k}$

$-35 = -50e^{-2k}$

$$\frac{7}{10} = e^{-2k}$$

$$\ln\left(\frac{7}{10}\right) = -2k \quad \text{or} \quad k = -\frac{1}{2}\ln\left(\frac{7}{10}\right) = 0.1783$$

Solution: $T = 70 - 50e^{-0.1783t}$

Now find t such that $T = 69°$:

$$69 = 70 - 50e^{-0.1783t}$$

$$\frac{1}{50} = e^{-0.1783t}$$

$$\ln\left(\frac{1}{50}\right) = -0.1783t$$

$$t = \frac{\ln(1/50)}{-0.1783} = 21.9 \text{ min}$$

17. From the equation

$$m\frac{dv}{dt} = F - kv \qquad\qquad F = mg$$

we get

$$10\frac{dv}{dt} = 100 - 0.2v$$

$$\frac{dv}{dt} + \frac{0.2}{10}v = 10$$

$$\frac{dv}{dt} + 0.02v = 10 \qquad\qquad I.F. = e^{0.02t}$$

$$\frac{d}{dt}(ve^{0.02t}) = 10e^{0.02t}$$

$$ve^{0.02t} = \frac{10}{0.02}e^{0.02t} + c = 500e^{0.02t} + c$$

$$v = 500 + ce^{-0.02t}$$

Initial condition: if $t = 0$, $v = 0$, or $0 = 500 + c$, and $c = -500$.

$$v = 500 - 500e^{-0.02t}$$

$$= 500(1 - e^{-0.02t})$$

If $t = 10$, then

$$v = 500(1 - e^{-0.2}) = 91 \text{ m/s}$$

Also,

$$\lim_{t \to \infty} v = \lim_{t \to \infty} 500(1 - e^{-0.02t}) = 500 \text{ m/s}$$

21. $\frac{dx}{dt} = kx$

$$\frac{dx}{dt} - kx = 0 \qquad\qquad\qquad \text{I.F.} = e^{-kt}$$

$$\frac{d}{dt}(xe^{-kt}) = 0$$

$$xe^{-kt} = c \qquad \text{and} \qquad x = ce^{kt}$$

If $x = x_0$ when $t = 0$, then $c = x_0$

$$x = x_0 e^{kt}$$

From the given condition, $x = \frac{3}{4}x_0$ when $t = 10$, we have

$$\frac{3}{4}x_0 = x_0 e^{10k}$$

$$\ln\left(\frac{3}{4}\right) = 10k \qquad \text{or} \qquad k = \frac{\ln(3/4)}{10} = -0.02877$$

Thus

$$x = x_0 e^{-0.02877t}$$

If $x = \frac{1}{10}x_0$, then   one-tenth unconverted

$$\frac{1}{10}x_0 = x_0 e^{-0.02877t}$$

$$\ln\left(\frac{1}{10}\right) = -0.02877t$$

$$t = \frac{\ln(1/10)}{-0.02877} = 80 \text{ s}$$

25. $y^2 = 4px$. Differentiating implicitly, we get

$$2y\frac{dy}{dx} = 4p$$

From the first equation, $4p = \frac{y^2}{x}$. Substituting in the second equation, we get

$$2y \frac{dy}{dx} = \frac{y^2}{x} \quad \text{or} \quad \frac{dy}{dx} = \frac{y}{2x}$$

So the orthogonal trajectories satisfy the condition

$$\frac{dy}{dx} = -\frac{2x}{y}$$

$$y\,dy = -2x\,dx$$

$$\frac{y^2}{2} = -x^2 + k_1$$

Thus

$$2x^2 + y^2 = k$$

29. $xy = c$ $\qquad\qquad\qquad\qquad\qquad$ given family

$$y = \frac{c}{x}$$

$$\frac{dy}{dx} = -\frac{c}{x^2}$$

Substituting $c = xy$, we get

$$\frac{dy}{dx} = -\frac{xy}{x^2} = -\frac{y}{x}$$

The orthogonal family therefore satisfies the condition

$$\frac{dy}{dx} = \frac{x}{y}$$

$$y\,dy = x\,dx$$

$$\tfrac{1}{2}y^2 = \tfrac{1}{2}x^2 + k_1$$

$$y^2 = x^2 + 2k_1$$

$$y^2 - x^2 = k$$

REVIEW EXERCISES FOR CHAPTER 11

1. $y' = x - y$

$$\frac{dy}{dx} + y = x \qquad\qquad\qquad \text{I.F.} = e^x$$

$$\frac{d}{dx}(ye^x) = xe^x$$

$$ye^x = \int xe^x\,dx$$

$$\qquad\qquad\qquad\qquad\qquad u = x \qquad dv = e^x\,dx$$

$$\qquad\qquad\qquad\qquad\qquad du = dx \qquad v = e^x$$

$$ye^x = xe^x - \int e^x\,dx = xe^x - e^x + c$$

$$y = x - 1 + ce^{-x}$$

5. $(1 + y^2)dx + (x^2y + y)dy = 0$

       $(1 + y^2)dx + y(x^2 + 1)dy = 0$

$$\frac{dx}{1 + x^2} + \frac{y\,dy}{1 + y^2} = 0 \qquad\qquad u = 1 + y^2; \; du = 2y\,dy$$

$$\frac{2\,dx}{1 + x^2} + \frac{2y\,dy}{1 + y^2} = 0 \qquad\qquad \text{multiplying by 2}$$

$$\int \frac{2\,dx}{1 + x^2} + \int \frac{2y\,dy}{1 + y^2} = c$$

$$2 \text{ Arctan } x + \ln(1 + y^2) = c$$

9. $(x^4 + 2y)dx - x\,dy = 0$

       $-x\,dy + 2y\,dx = -x^4\,dx$

$$\frac{dy}{dx} - \left(\frac{2}{x}\right)y = x^3 \qquad\qquad \text{I.F.} = e^{-\int (2/x)dx} = e^{-2\ln x} = e^{\ln x^{-2}} = x^{-2}$$

$$\frac{d}{dx}(x^{-2}y) = x^{-2}x^3 = x$$

$$x^{-2}y = \tfrac{1}{2}x^2 + c$$

$$y = \tfrac{1}{2}x^4 + cx^2$$

13.   **As a linear equation:**

$$\frac{dy}{dx} + (\sec x)y = 0 \qquad\qquad \text{I.F.} = e^{\int \sec x\,dx} = e^{\ln(\sec x + \tan x)} = \sec x + \tan x$$

$$\frac{d}{dx}[y(\sec x + \tan x)] = 0$$

$$y(\sec x + \tan x) = c$$

Separation of variables:

$$dy + y \sec x\,dx = 0$$

$$\frac{dy}{y} + \sec x\,dx = 0$$

$$\ln y + \ln(\sec x + \tan x) = \ln c \qquad\qquad \text{integrating}$$

$$\ln y(\sec x + \tan x) = \ln c$$

$$y(\sec x + \tan x) = c$$

Since $y = 2$ when $x = \frac{\pi}{4}$, we get

$$2\left[\sec\left(\frac{\pi}{4}\right) + \tan\left(\frac{\pi}{4}\right)\right] = c$$

$$2(\sqrt{2} + 1) = c$$

Hence

$$y(\sec x + \tan x) = 2(\sqrt{2} + 1)$$

17. Since $M_T = 65°F$, the equation is

$$\frac{dT}{dt} = -k(T - 65)$$

$$\frac{dT}{dt} + kT = 65k \qquad\qquad I.F. = e^{kt}$$

$$\frac{d}{dt}(Te^{kt}) = 65ke^{kt}$$

$$Te^{kt} = 65e^{kt} + c$$

$$T = 65 + ce^{-kt}$$

If $t = 0$, $T = T_0$ (unknown initial temperature). Thus

$T_0 = 65 + c$ or $c = T_0 - 65$

$T = 65 + (T_0 - 65)e^{-kt}$

If $t = 15$, then $T = 0$, and if $t = 30$, then $T = 20$:

$$0 = 65 + (T_0 - 65)e^{-15k} \qquad\qquad (1)$$

$$\underline{20 = 65 + (T_0 - 65)e^{-30k}}$$

$$-65e^{15k} = T_0 - 65$$

$$\underline{-45e^{30k} = T_0 - 65}$$

$-65e^{15k} + 45e^{30k} = 0$ (subtracting)

$-65 \quad\;\; + 45e^{15k} = 0$ (dividing by e^{15k})

$$e^{15k} = \frac{65}{45}$$

$$15k = \ln\left(\frac{65}{45}\right)$$

$$k = \frac{1}{15}\ln\left(\frac{65}{45}\right) = 0.0245$$

Substituting in (1), we get

$$0 = 65 + (T_0 - 65)e^{-15(0.0245)}$$

$$T_0 - 65 = -65e^{15(0.0245)}$$

$$T_0 = -29°F$$

21. Differentiating implicitly, we get

$2x - 4y\dfrac{dy}{dx} = 0$ or $\dfrac{dy}{dx} = \dfrac{x}{2y}$

Condition for orthogonal trajectories:

$$\frac{dy}{dx} = -\frac{2y}{x}$$

$$\frac{dy}{y} = -\frac{2\,dx}{x}$$

$$\ln y = -2\ln x + \ln k$$

$$\ln y + 2\ln x = \ln k$$

$$\ln y + \ln x^2 = \ln k, \text{ whence } x^2 y = k$$

CHAPTER 12   HIGHER-ORDER LINEAR DIFFERENTIAL EQUATIONS

1. $m^2 - 13m + 42 = 0$

$(m - 6)(m - 7) = 0$

$$m = 6, 7$$

$y = c_1 e^{6x} + c_2 e^{7x}$

5. $4m^2 + 7m - 2 = 0$

$(4m - 1)(m + 2) = 0$

$$m = \frac{1}{4}, -2$$

$y = c_1 e^{x/4} + c_2 e^{-2x}$

9. $2m^2 - 3m + 1 = 0$

$(2m - 1)(m - 1) = 0$

$$m = 1, \frac{1}{2}$$

$y = c_1 e^x + c_2 e^{x/2}$

13. $m^2 - m - 2 = 0$

$(m - 2)(m + 1) = 0$

$$m = 2, -1$$

$y = c_1 e^{2x} + c_2 e^{-x}$

Substituting $(0,0)$ and $(1,1)$, respectively, we get

(1) $0 = c_1 + c_2$ $\qquad\qquad$ $y = 0, x = 0$

(2) $1 = c_1 e^2 + c_2 e^{-1}$ $\qquad\quad$ $y = 1, x = 1$

From the first equation, $c_1 = -c_2$. Substituting in the second equation, we get

$$1 = -c_2 e^2 + c_2 e^{-1}$$
$$1 = c_2(-e^2 + e^{-1})$$

$$c_2 = \frac{1}{e^{-1} - e^2} = \frac{e}{1 - e^3}$$

Thus

$$c_1 = -\frac{e}{1 - e^3}$$

So the solution becomes

$$y = -\frac{e}{1 - e^3}e^{2x} + \frac{e}{1 - e^3}e^{-x}$$

$$= \frac{e}{1 - e^3}(e^{-x} - e^{2x})$$

17. $m^3 - m^2 - 4m - 2 = 0$

possible rational roots: $\pm 1, \pm 2$

```
1 - 1 - 4 - 2) -1
   - 1 + 2 + 2
  1 - 2 - 2 + 0          x = -1 is a root
```

$m^2 - 2m - 2 = 0$

$$m = \frac{-(-2) \pm \sqrt{(-2)^2 - 4(1)(-2)}}{2 \cdot 1}$$

$$= \frac{2 \pm \sqrt{4 + 8}}{2} = \frac{2 \pm 2\sqrt{3}}{2} = 1 \pm \sqrt{3}$$

So the roots are: $-1, 1 \pm \sqrt{3}$

$$y = c_1 e^{-x} + c_2 e^{(1+\sqrt{3})x} + c_2 e^{(1-\sqrt{3})x}$$

21. $m^2 - 2m - 2 = 0$

$$m = \frac{2 \pm \sqrt{4 + 8}}{2} = \frac{2 \pm 2\sqrt{3}}{2} = 1 \pm \sqrt{3}$$

$$y = c_1 e^{(1+\sqrt{3})x} + c_2 e^{(1-\sqrt{3})x}$$

$$y = c_1 e^x e^{\sqrt{3}x} + c_2 e^x e^{-\sqrt{3}x}$$

$$y = e^x\left(c_1 e^{\sqrt{3}x} + c_2 e^{-\sqrt{3}x}\right)$$

25. $m^2 + 6m - 6 = 0$

$$m = \frac{-6 \pm \sqrt{36 + 24}}{2} = \frac{-6 \pm \sqrt{4 \cdot 15}}{2} = \frac{-6 \pm 2\sqrt{15}}{2}$$

$$= -3 \pm \sqrt{15}$$

$$y = c_1 e^{(-3+\sqrt{15})x} + c_2 e^{(-3-\sqrt{15})x}$$

$$= c_1 e^{-3x} e^{\sqrt{15}x} + c_2 e^{-3x} e^{-\sqrt{15}x}$$

$$= e^{-3x}(c_1 e^{\sqrt{15}x} + c_2 e^{-\sqrt{15}x})$$

29. $3m^2 - m - 2 = 0$

$(3m + 2)(m - 1) = 0$

$$m = 1, -\frac{2}{3}$$

$$y = c_1 e^x + c_2 e^{-(2/3)x}$$

Section 12.2

1. $m^2 + 6m + 9 = 0$

    $(m + 3)^2 = 0$

    $m = -3, -3$ (repeating root)

    $y = c_1 e^{-3x} + c_2 x e^{-3x}$

5. $9m^2 + 12m + 4 = 0$

    $(3m + 2)^2 = 0$

    $m = -\frac{2}{3}, -\frac{2}{3}$

    $y = c_1 e^{-(2/3)x} + c_2 x e^{-(2/3)x}$

9. $m^2 - 4m + 5 = 0$

    $$m = \frac{4 \pm \sqrt{16 - 20}}{2} = \frac{4 \pm \sqrt{-4}}{2} = \frac{4 \pm 2j}{2} = 2 \pm j$$

    $y = e^{2x}(c_1 \cos x + c_2 \sin x)$

13. $2m^2 - 2m + 1 = 0$

    $$m = \frac{2 \pm \sqrt{4 - 8}}{4} = \frac{2 \pm \sqrt{-4}}{4} = \frac{2 \pm 2j}{4} = \frac{1}{2} \pm \frac{1}{2}j$$

    $y = e^{(1/2)x}\left(c_1 \cos \frac{1}{2}x + c_2 \sin \frac{1}{2}x\right)$

17. $m^2 - 6m + 9 = 0$

    $(m - 3)^2 = 0$

    $m = 3, 3$

    $y = c_1 e^{3x} + c_2 x e^{3x}$

21. $m^2 + m + 2 = 0$

    $$m = \frac{-1 \pm \sqrt{1 - 8}}{2} = \frac{-1 \pm \sqrt{7}j}{2} = -\frac{1}{2} \pm \frac{\sqrt{7}}{2}j$$

    $y = e^{-x/2}\left(c_1 \cos \frac{\sqrt{7}}{2}x + c_2 \sin \frac{\sqrt{7}}{2}x\right)$

25. $2m^2 - 4m + 5 = 0$

    $$m = \frac{4 \pm \sqrt{16 - 40}}{4} = \frac{4 \pm \sqrt{-24}}{4} = \frac{4 \pm \sqrt{4 \cdot (-6)}}{4}$$

234

$$= \frac{4 \pm 2\sqrt{6}\,j}{4} = 1 \pm \tfrac{1}{2}\sqrt{6}\,j$$

$$y = e^x\left(c_1 \cos \tfrac{1}{2}\sqrt{6}x + c_2 \sin \tfrac{1}{2}\sqrt{6}x\right)$$

29. $\quad m^2 - 100 = 0$

$$m = \pm 10$$

$$y = c_1 e^{10x} + c_2 e^{-10x}$$

33. $\qquad 3m^3 - 2m^2 + m = 0$

$$m(3m^2 - 2m + 1) = 0$$

$$m = 0,$$

$$m = \frac{2 \pm \sqrt{4 - 12}}{6} = \frac{2 \pm \sqrt{2(4)(-1)}}{6} = \frac{2 \pm 2\sqrt{2}\,j}{6}$$

$$= \tfrac{1}{3} \pm \tfrac{1}{3}\sqrt{2}\,j$$

$$y = c_1 + e^{x/3}\left(c_2 \cos \tfrac{\sqrt{2}}{3}x + c_3 \sin \tfrac{\sqrt{2}}{3}x\right)$$

37. $\quad m^2 + 4 = 0$

$$m^2 = -4$$

$$m = \pm 2j$$

$$y = e^{0x}(c_1 \cos 2x + c_2 \sin 2x) = c_1 \cos 2x + c_2 \sin 2x$$

41. $\quad m^3 - 2m^2 + 2m - 1 = 0$

The only possible rational roots are ± 1. By inspection, $m = 1$ is a root and $m - 1$ a factor:

```
1 - 2 + 2 - 1) 1
      1 - 1 + 1
1 - 1 + 1 + 0          m = 1 is a root
```

So the factor is $m^2 - m + 1$, and the remaining roots are

$$m = \frac{1 \pm \sqrt{1 - 4}}{2} = \tfrac{1}{2} \pm \tfrac{1}{2}\sqrt{3}\,j$$

$$y = c_1 e^x + e^{x/2}\left(c_2 \cos \tfrac{\sqrt{3}}{2}x + c_3 \sin \tfrac{\sqrt{3}}{2}x\right)$$

45. $\quad m^2 + 1 = 0$

$$m = \pm j$$

$$y = c_1 \cos x + c_2 \sin x$$

Substituting $(0,0)$ and $\left(\tfrac{\pi}{2},1\right)$, respectively, we get

(1) $0 = c_1 \cdot 1 + c_2 \cdot 0$ or $c_1 = 0$ $y = 0,\ x = 0$

(2) $1 = c_1 \cdot 0 + c_2 \cdot 1$ or $c_2 = 1$ $y = 1,\ x = \tfrac{\pi}{2}$

Thus $y = \sin x$ is the solution.

49.　$m^4 + 18m^2 + 81 = 0$

$$(m^2 + 9)^2 = (m^2 + 9)(m^2 + 9) = 0$$

$m = 3j, \ 3j, \ -3j, \ -3j$

$y = c_1 e^{3jx} + c_2 x e^{3jx} + c_3 e^{-3jx} + c_4 x e^{-3jx}$

$\quad = c_1(\cos 3x + j \sin 3x) + c_2 x(\cos 3x + j \sin 3x)$

$\qquad\qquad + c_3(\cos 3x - j \sin 3x) + c_4 x(\cos 3x - j \sin 3x)$

$\quad = (c_1 + c_3)\cos 3x + j(c_1 - c_3)\sin 3x$

$\qquad\qquad + (c_2 + c_4)x \cos 3x + j(c_2 - c_4)x \sin 3x$

So y has the following form:

$y = c_1 \cos 3x + c_2 \sin 3x + c_3 x \cos 3x + c_4 x \sin 3x$

Section 12.3

1.　$(D^2 - 6D + 9)y = e^x$

$y_c:$　$m^2 - 6m + 9 = 0$

$\qquad\qquad (m - 3)^2 = 0$

$\qquad\qquad\qquad m = 3, \ 3$

$y_c = c_1 e^{3x} + c_2 x e^{3x}$

$y_p:$　$y_p = Ae^x; \ y_p' = Ae^x; \ y_p'' = Ae^x$

Substituting in $(D^2 - 6D + 9)y = e^x$:

$\quad y_p'' - 6y_p' + 9y_p = e^x$

$Ae^x - 6Ae^x + 9Ae^x = e^x$

$\qquad\qquad 4Ae^x = e^x$

$\qquad\qquad\quad A = \frac{1}{4}$

So $y_p = Ae^x = \frac{1}{4}e^x$

and

$y = y_c + y_p = c_1 e^{3x} + c_2 x e^{3x} + \frac{1}{4}e^x$

5.　$(D^2 - D - 2)y = 2x^2$

$y_c:$　$m^2 - m - 2 = (m - 2)(m + 1) = 0$

$\qquad\qquad\qquad\qquad\qquad m = 2, \ -1$

$\qquad y_c = c_1 e^{2x} + c_2 e^{-x}$

$y_p:$　$y_p = Ax^2 + Bx + C, \ y_p' = 2Ax + B, \ y_p'' = 2A$

Substituting in

$\qquad (D^2 - D - 2)y = 2x^2$

we get

$$2A - (2Ax + B) - 2(Ax^2 + Bx + C) = 2x^2$$

$$2A - 2Ax - B - 2Ax^2 - 2Bx - 2C = 2x^2$$

$$-2Ax^2 + (-2A - 2B)x + (2A - B - 2C) = 2x^2 + 0x + 0$$

$$-2A = 2 \qquad x^2\text{-coefficients}$$

$$-2A - 2B = 0 \qquad x\text{-coefficients}$$

$$2A - B - 2C = 0 \qquad \text{constants}$$

We obtain: $A = -1$, $B = 1$, and $C = -\frac{3}{2}$.

So

$$y_p = Ax^2 + Bx + C = -x^2 + x - \frac{3}{2}$$

and

$$y = y_c + y_p = c_1 e^{2x} + c_2 e^{-x} - x^2 + x - \frac{3}{2}$$

9. y_c: $\qquad m^2 + 4m + 3 = 0$

$$(m + 1)(m + 3) = 0$$

$$m = -1, -3$$

$$y_c = c_1 e^{-x} + c_2 e^{-3x}$$

y_p: $y_p = A + Be^x$; $y_p' = Be^x$; $y_p'' = Be^x$

Substituting in the equation:

$$y'' + 4y' + 3y = 6 + e^x$$

$$Be^x + 4Be^x + 3(A + Be^x) = 6 + e^x$$

$$8Be^x + 3A = 6 + e^x$$

$$A = 2, \ B = \frac{1}{8}$$

So

$$y_p = 2 + \frac{1}{8}e^x$$

and

$$y = c_1 e^{-x} + c_2 e^{-3x} + \frac{1}{8}e^x + 2$$

13. y_c: $\qquad m^2 + 5m + 6 = 0$

$$(m + 2)(m + 3) = 0$$

$$m = -2, -3$$

$$y_c = c_1 e^{-2x} + c_2 e^{-3x}$$

y_p: $y_p = A \cos x + B \sin x$

$$y_p' = -A \sin x + B \cos x$$

$$y_p'' = -A \cos x - B \sin x$$

Substituting in the equation

$$(D^2 + 5D + 6)y = 4 \cos x + 6 \sin x$$

we get

$$(-A \cos x - B \sin x) + 5(-A \sin x + B \cos x)$$
$$+ 6(A \cos x + B \sin x) = 4 \cos x + 6 \sin x$$

$$(5A + 5B)\cos x + (-5A + 5B)\sin x = 4 \cos x + 6 \sin x$$

Comparing coefficients:

$$\begin{array}{r} 5A + 5B = 4 \\ -5A + 5B = 6 \\ \hline 10B = 10, \ B = 1 \end{array}$$

$$5A + 5 = 4, \ A = -\frac{1}{5}$$

So

$$y_p = -\frac{1}{5} \cos x + \sin x$$

$$y = c_1 e^{-2x} + c_2 e^{-3x} - \frac{1}{5} \cos x + \sin x$$

17. y_c: $m^3 - 2m^2 - m + 2 = 0$

Possible rational roots: $\pm 1, \pm 2$; one root is $m = 1$.

$$\begin{array}{r} 1 - 2 - 1 + 2)\underline{\ 1\ \ } \\ \underline{1 - 1 - 2} \\ 1 - 1 - 2 + 0 \end{array}$$

Since the resulting factor is $m^2 - m - 2 = (m - 2)(m + 1)$, the other roots are 2 and -1.

$$y_c = c_1 e^x + c_2 e^{-x} + c_3 e^{2x}$$

y_p: $y_p = Ae^{3x}$; $y_p' = 3Ae^{3x}$; $y_p'' = 9Ae^{3x}$; $y_p''' = 27Ae^{3x}$

Substituting in the equation:

$$27Ae^{3x} - 2(9Ae^{3x}) - 3Ae^{3x} + 2Ae^{3x} = 8e^{3x}$$
$$8A = 8, \ A = 1 \quad \text{and} \quad y_p = e^{3x}$$

$$y = c_1 e^x + c_2 e^{-x} + c_3 e^{2x} + e^{3x}$$

21. y_c: $m^2 + 9 = 0$

$$m = \pm 3j$$

$$y_c = c_1 \cos 3x + c_2 \sin 3x$$

y_p: $y_p = Ae^{3x}$; $y_p' = 3Ae^{3x}$; $y_p'' = 9Ae^{3x}$

Substituting in the equation:

$$9Ae^{3x} + 9Ae^{3x} = 9e^{3x}$$
$$18A = 9, \ A = \frac{1}{2} \quad \text{and} \quad y_p = \frac{1}{2}e^{3x}$$

$y = c_1 \cos 3x + c_2 \sin 3x + \frac{1}{2}e^{3x}$

$Dy = -3c_1 \sin 3x + 3c_2 \cos 3x + \frac{3}{2}e^{3x}$

Substituting given conditions, we have

(1) $\quad 1 = \quad c_1 \cdot 1 + \quad c_2 \cdot 0 + \frac{1}{2} \qquad\qquad y = 1, \; x = 0$

(2) $\quad \frac{3}{2} = -3c_1 \cdot 0 + 3c_2 \cdot 1 + \frac{3}{2} \qquad\qquad Dy = \frac{3}{2}, \; x = 0$

$\qquad\qquad c_1 = \frac{1}{2}, \; c_2 = 0$

The solution is

$y = \frac{1}{2}\cos 3x + \frac{1}{2}e^{3x} = \frac{1}{2}(\cos 3x + e^{3x})$

25. $\quad (D^2 + 2D + 5)y = 10 \cos x$

$\quad y_c: \quad m^2 + 2m + 5 = 0$

$$m = \frac{-2 \pm \sqrt{4 - 20}}{2} = \frac{-2 \pm 4j}{2} = -1 \pm 2j$$

$y_c = e^{-x}(c_1 \cos 2x + c_2 \sin 2x)$

$\quad y_p: \quad y_p = A \cos x + B \sin x$

$\qquad\qquad y_p' = -A \sin x + B \cos x$

$\qquad\qquad y_p'' = -A \cos x - B \sin x$

Substituting in the given equation:

$(-A \cos x - B \sin x) + 2(-A \sin x + B \cos x)$

$\qquad\qquad + 5(A \cos x + B \sin x) = 10 \cos x$

$(4A + 2B)\cos x + (-2A + 4B)\sin x = 10 \cos x$

$(4A + 2B)\cos x + (-2A + 4B)\sin x = 10 \cos x + 0 \sin x$

$\quad\begin{array}{l} 4A + 2B = 10 \\ \underline{-2A + 4B = \;\; 0} \\ 4A + 2B = 10 \\ \underline{-4A + 8B = \;\; 0} \\ \qquad 10B = 10, \; B = 1 \end{array}$

$-2A + \;\; 4 = \;\; 0, \; A = 2 \qquad\quad y_p = 2 \cos x + \sin x$

General solution:

$\quad y = e^{-x}(c_1 \cos 2x + c_2 \sin 2x) + 2 \cos x + \sin x$

$Dy = e^{-x}(-2c_1 \sin 2x + 2c_2 \cos 2x) - e^{-x}(c_1 \cos 2x + c_2 \sin 2x)$

$\qquad\qquad\qquad\qquad\qquad\qquad - 2 \sin x + \cos x$

Now substitute the given conditions:

(1) $\quad 5 = c_1 + 2 \qquad\qquad x = 0, \; y = 5$

(2) $\quad 6 = 2c_2 - c_1 + 1 \qquad x = 0, \; Dy = 6$

Thus $c_1 = 3$ and $c_2 = 4$, so that the solution is
$$y = e^{-x}(3 \cos 2x + 4 \sin 2x) + 2 \cos x + \sin x$$

29. $(D^2 - D + 2)y = \cos x$

$y_p = A \cos x + B \sin x$
$y_p' = -A \sin x + B \cos x$
$y_p'' = -A \cos x - B \sin x$

Substituting:

$(-A \cos x - B \sin x) - (-A \sin x + B \cos x)$
$\qquad + 2(A \cos x + B \sin x) = \cos x$
$(A - B)\cos x + (A + B)\sin x = \cos x$

$\begin{array}{l} A - B = 1 \\ \underline{A + B = 0} \\ 2A \qquad = 1, \ A = \frac{1}{2} \\ \qquad B = -A, \ B = -\frac{1}{2} \end{array}$

$y_p = \frac{1}{2} \cos x - \frac{1}{2} \sin x$

33. $(D^2 + 2)y = 2e^{2x} + 2$

$y_p = Ae^{2x} + B; \ y_p' = 2Ae^{2x}; \ y_p'' = 4Ae^{2x}$

Substituting:

$4Ae^{2x} + 2(Ae^{2x} + B) = 2e^{2x} + 2$
$\qquad 6Ae^{2x} + 2B = 2e^{2x} + 2$
$\qquad\qquad A = \frac{1}{3}, \ B = 1$

$y_p = \frac{1}{3}e^{2x} + 1$

37. $(D^2 + 3)y = 2x + \cos x$

$y_p = Ax + B + C \cos x + D \sin x$
$y_p' = A - C \sin x + D \cos x$
$y_p'' = -C \cos x - D \sin x$

Substituting:

$(-C \cos x - D \sin x) + 3(Ax + B + C \cos x + D \sin x)$
$\qquad\qquad\qquad\qquad\qquad = 2x + \cos x$
$3Ax + 3B + 2C \cos x + 2D \sin x = 2x + \cos x$

$A = \frac{2}{3}, \ B = 0, \ C = \frac{1}{2}, \ D = 0$

$y_p = \frac{2}{3}x + \frac{1}{2} \cos x$

41. $(D^2 - 4)y = 8e^{2x}$

y_c: $m^2 - 4 = 0$, $m = \pm 2$

$y_c = c_1 e^{-2x} + c_2 e^{2x}$

Since the right side of the given equation, $8e^{2x}$, is a term in y_c, we need the annihilator. (In other words, y_p is not of the form Ae^{2x}.)

Right side: $8e^{2x}$.

$$m' = 2$$

$$m' - 2 = 0$$

Annihilator: $D - 2$

Applying the annihilator to the equation, we get

$$(D - 2)(D^2 - 4)y = (D - 2)(8e^{2x}) = 0$$

$$(m - 2)(m^2 - 4) = 0$$

$$(m - 2)(m - 2)(m + 2) = 0$$

$$m = 2, 2, -2$$

So y has the form

$$y = c_1 e^{-2x} + c_2 e^{2x} + c_3 x e^{2x}$$

Since $y_c = c_1 e^{-2x} + c_2 e^{2x}$, we conclude that

$$y_p = Axe^{2x}$$

$$y_p' = 2Axe^{2x} + Ae^{2x}$$

$$y_p'' = 4Axe^{2x} + 2Ae^{2x} + 2Ae^{2x} = 4Axe^{2x} + 4Ae^{2x}$$

Substituting in the given equation, we get

$$(4Axe^{2x} + 4Ae^{2x}) - 4Axe^{2x} = 8e^{2x}$$

$$4Ae^{2x} = 8e^{2x}$$

$$A = 2$$

$$y = y_c + y_p = c_1 e^{-2x} + c_2 e^{2x} + 2xe^{2x}$$

45. $(D^2 + 3D - 28)y = 11e^{4x}$

y_c: $m^2 + 3m - 28 = (m + 7)(m - 4) = 0$

$$m = 4, -7$$

$$y_c = c_1 e^{-7x} + c_2 e^{4x}$$

(Note that the right side, $11e^{4x}$, is one of the terms in y_c.)

y_p: Right side $= 11e^{4x}$

$$m' = 4$$

$$m' - 4 = 0$$

Annihilator: D - 4

Applying the annihilator to the equation, we get

$$(D - 4)(D^2 + 3D - 28)y = (D - 4)(11e^{4x}) = 0$$

$$(m - 4)(m - 4)(m + 7) = 0$$

$$m = 4, 4, -7$$

So y has the following form:

$$y = c_1e^{-7x} + c_2e^{4x} + c_3xe^{4x}$$

Since $y_c = c_1e^{-7x} + c_2e^{4x}$, we conclude that

$$y_p = Axe^{4x}$$

$$y_p' = 4Axe^{4x} + Ae^{4x}$$

$$y_p'' = 16Axe^{4x} + 4Ae^{4x} + 4Ae^{4x} = 16Axe^{4x} + 8Ae^{4x}$$

Substituting in the given equation,

$$(16Axe^{4x} + 8Ae^{4x}) + 3(4Axe^{4x} + Ae^{4x}) - 28Axe^{4x} = 11e^{4x}$$

$$16Axe^{4x} + 8Ae^{4x} + 12Axe^{4x} + 3Ae^{4x} - 28Axe^{4x} = 11e^{4x}$$

$$11Ae^{4x} = 11e^{4x}$$

$$A = 1$$

So

$$y_p = xe^{4x}$$

and

$$y = y_c + y_p = c_1e^{-7x} + c_2e^{4x} + xe^{4x}$$

49. $(D^2 - 4)y = 4 + e^{2x}$

y_c: $m^2 - 4 = 0$, $m = \pm 2$

$$y_c = c_1e^{-2x} + c_2e^{2x}$$

y_p: Right side $= 4 + e^{2x}$

$$m' = 0, 2$$

$$m'(m' - 2) = 0$$

Annihilator: $D(D - 2)$

Applying the annihilator to the equation:

$$D(D - 2)(D^2 - 4)y = D(D - 2)(4 + e^{2x}) = 0$$

$$m(m - 2)(m^2 - 4) = 0$$

$$m = 0, 2, 2, -2$$

So y has the form

$$y = c_1e^{-2x} + c_2e^{2x} + c_3xe^{2x} + c_4e^{0x}$$

Since $y_c = c_1e^{-2x} + c_2e^{2x}$, we conclude that

$$y_p = Axe^{2x} + B$$
$$y_p' = 2Axe^{2x} + Ae^{2x}$$
$$y_p'' = 4Axe^{2x} + 2Ae^{2x} + 2Ae^{2x} = 4Axe^{2x} + 4Ae^{2x}$$

Substituting in the given equation,

$$(4Axe^{2x} + 4Ae^{2x}) - 4(Axe^{2x} + B) = 4 + e^{2x}$$
$$4Axe^{2x} + 4Ae^{2x} - 4Axe^{2x} - 4B = 4 + e^{2x}$$
$$-4B + 4Ae^{2x} = 4 + e^{2x}$$
$$-4B = 4$$
$$4A = 1$$

So $A = \frac{1}{4}$ and $B = -1$ and $y_p = \frac{1}{4}xe^{2x} - 1$

$$y = y_c + y_p = c_1e^{-2x} + c_2e^{2x} + \frac{1}{4}xe^{2x} - 1$$

Section 12.4

1. By Hooke's law

$$F = kx$$
$$4 = k \cdot \frac{1}{2} \quad \text{or} \quad k = 8 \qquad \qquad \text{spring constant}$$

$$\text{mass} = \frac{4}{32} = \frac{1}{8} \text{ slug} \qquad \qquad \text{mass}$$

$$m\frac{d^2x}{dt^2} + kx = 0 \qquad \qquad b = 0$$
$$\frac{1}{8}\frac{d^2x}{dt^2} + 8x = 0$$
$$\frac{d^2x}{dt^2} + 64x = 0 \qquad \qquad \text{equation}$$

Conditions:

(1) If $t = 0$, $x = \frac{1}{4}$ ft $\qquad \qquad$ initial position

(2) If $t = 0$, $\frac{dx}{dt} = 0$ $\qquad \qquad$ initial velocity

Auxiliary equation:

$$m^2 + 64 = 0, \quad m = \pm 8j$$

Thus

$$x(t) = c_1 \cos 8t + c_2 \sin 8t$$
$$x'(t) = -8c_1 \sin 8t + 8c_2 \cos 8t$$

From initial conditions:

$$\tfrac{1}{4} = c_1 + c_2(0) \qquad \text{or} \qquad c_1 = \tfrac{1}{4}$$
$$0 = 0 + 8c_2 \qquad \text{or} \qquad c_2 = 0$$

Substituting c_1 and c_2, we get
$$x(t) = \tfrac{1}{4} \cos 8t$$

5. By Hooke's law,

$$F = kx$$
$$12 = k \cdot 2 \quad \text{or} \quad k = 6 \qquad\qquad \text{spring constant}$$

mass: $\tfrac{12}{32} = \tfrac{3}{8}$ slug $\qquad\qquad$ mass

$$\tfrac{3}{8}\frac{d^2x}{dt^2} + 6x = 0 \qquad\qquad b = 0$$
$$\frac{d^2x}{dt^2} + 16x = 0 \qquad\qquad \text{equation}$$

Initial conditions:

(1) If $t = 0$, $x = \tfrac{8}{12} = \tfrac{2}{3}$ ft $\qquad\qquad$ initial position

(2) If $t = 0$, $\frac{dx}{dt} = 3$ ft/s $\qquad\qquad$ initial velocity

Auxiliary equation:
$$m^2 + 16 = 0 \quad \text{or} \quad m = \pm 4j$$
Thus
$$x(t) = c_1 \cos 4t + c_2 \sin 4t$$
$$x'(t) = -4c_1 \sin 4t + 4c_2 \cos 4t$$

From initial conditions:

$$\tfrac{2}{3} = c_1 + c_2(0) \qquad \text{or} \qquad c_1 = \tfrac{2}{3}$$
$$3 = 0 + 4c_2 \qquad \text{or} \qquad c_2 = \tfrac{3}{4}$$

Substituting c_1 and c_2:
$$x(t) = \tfrac{2}{3} \cos 4t + \tfrac{3}{4} \sin 4t$$

9. From Hooke's law:

$$F = kx$$
$$4 = k \cdot 2 \quad \text{or} \quad k = 2 \qquad\qquad \text{spring constant}$$

mass: $\tfrac{4}{32} = \tfrac{1}{8}$ slug $\qquad\qquad$ mass

$$\tfrac{1}{8}\frac{d^2x}{dt^2} + \tfrac{1}{2}\frac{dx}{dt} + 2x = 0 \qquad\qquad b = \tfrac{1}{2}$$

244

$$\frac{d^2x}{dt^2} + 4\frac{dx}{dt} + 16x = 0 \qquad \text{equation}$$

Initial conditions:

1. If $t = 0$, $x = \frac{1}{2}$ ft   initial position

2. If $t = 0$, $\frac{dx}{dt} = 0$ initial velocity

Auxiliary equation:

$$m^2 + 4m + 16 = 0$$

$$m = \frac{-4 \pm \sqrt{16 - 64}}{2} = \frac{-4 \pm \sqrt{-48}}{2}$$

$$= \frac{-4 \pm 4\sqrt{3}\,j}{2} = -2 \pm 2\sqrt{3}\,j$$

Thus

$$x(t) = e^{-2t}(c_1 \cos 2\sqrt{3}\,t + c_2 \sin 2\sqrt{3}\,t)$$

By the product rule,

$$x'(t) = e^{-2t}(-2\sqrt{3}\,c_1 \sin 2\sqrt{3}\,t + 2\sqrt{3}\,c_2 \cos 2\sqrt{3}\,t)$$
$$-2e^{-2t}(c_1 \cos 2\sqrt{3}\,t + c_2 \sin 2\sqrt{3}\,t)$$

In the first equation, using $t = 0$ and $x = \frac{1}{2}$:

$$\tfrac{1}{2} = 1(c_1 + c_2 \cdot 0)$$

so that

$$c_1 = \tfrac{1}{2}$$

In the second equation, let $t = 0$ and $x'(t) = 0$:

$$0 = 1(0 + 2\sqrt{3}\,c_2) - 2(c_1 + 0)$$

so that $0 = 2\sqrt{3}\,c_2 - 2c_1$. Since $c_1 = \frac{1}{2}$,

$$2\sqrt{3}\,c_2 - 1 = 0$$

and

$$c_2 = \frac{1}{2\sqrt{3}} \qquad \cdot$$

It follows that

$$x(t) = e^{-2t}\left(\tfrac{1}{2} \cos 2\sqrt{3}\,t + \frac{1}{2\sqrt{3}} \sin 2\sqrt{3}\,t\right)$$

13. By Hooke's law

$$F = kx$$
$$19.6 = k \cdot 0.098 \qquad \text{or} \qquad k = 200$$

$$m\frac{d^2x}{dt^2} + kx = 0 \qquad\qquad\qquad b = 0$$

$$2.0\,\frac{d^2x}{dt^2} + 200x = 0$$

$$\frac{d^2x}{dt^2} + 100x = 0 \qquad\qquad\qquad \text{equation}$$

$$m^2 + 100 = 0 \qquad\qquad\qquad \text{auxiliary equation}$$

$$m = \pm 10j$$

$$x = c_1 \cos 10t + c_2 \sin 10t$$

$$\frac{dx}{dt} = -10c_1 \sin 10t + 10c_2 \cos 10t$$

From the description in the problem, we have the following initial conditions: if $t = 0$, then $x = 0.25$ and $\frac{dx}{dt} = 0$. It follows that

(1) $0.25 = c_1$

(2) $0 = 10c_2$, or $c_2 = 0$

Thus

$x(t) = 0.25 \cos 10t$

17. From Exercise 13, $k = 200$. We are also given that $m = 2.0$ kg, $b = 4$, and $f(t) = 20 \sin 5t$.

$$m \frac{d^2x}{dt^2} + b \frac{dx}{dt} + kx = f(t)$$

$$2.0 \frac{d^2x}{dt^2} + 4 \frac{dx}{dt} + 200x = 20 \sin 5t$$

and

$$\frac{d^2x}{dt^2} + 2 \frac{dx}{dt} + 100x = 10 \sin 5t \qquad\qquad (*)$$

x_c: $m^2 + 2m + 100 = 0$

$$m = \frac{-2 \pm \sqrt{4 - 400}}{2} = \frac{-2 \pm \sqrt{-396}}{2}$$

$$= -1 \pm \tfrac{1}{2}\sqrt{396}\,j = -1 \pm \sqrt{99}\,j$$

$x_c = e^{-t}(c_1 \cos \sqrt{99}\,t + c_2 \sin \sqrt{99}\,t)$

x_p: $x_p = A \cos 5t + B \sin 5t$

$x_p' = -5A \sin 5t + 5B \cos 5t$

$x_p'' = -25A \cos 5t - 25B \sin 5t$

Substituting in equation $(*)$:

$(-25A \cos 5t - 25B \sin 5t) + 2(-5A \sin 5t + 5B \cos 5t)$

$\qquad\qquad + 100(A \cos 5t + B \sin 5t) = 10 \sin 5t$

$(75A + 10B)\cos 5t + (-10A + 75B)\sin 5t = 10 \sin 5t$

  $75A + 10B = 0$

$-10A + 75B = 10$

$$A = \frac{\begin{vmatrix} 0 & 10 \\ 10 & 75 \end{vmatrix}}{\begin{vmatrix} 75 & 10 \\ -10 & 75 \end{vmatrix}} = \frac{0 - 100}{75^2 + 10^2} = \frac{-100}{5725} = -0.01747$$

$$B = \frac{\begin{vmatrix} 75 & 0 \\ -10 & 10 \end{vmatrix}}{\begin{vmatrix} 75 & 10 \\ -10 & 75 \end{vmatrix}} = \frac{750}{5725} = 0.131$$

General solution:

$$x(t) = e^{-t}(c_1 \cos \sqrt{99}\,t + c_2 \sin \sqrt{99}\,t)$$
$$- 0.01747 \cos 5t + 0.131 \sin 5t$$

$$x'(t) = e^{-t}(-\sqrt{99}\,c_1 \sin \sqrt{99}\,t + \sqrt{99}\,c_2 \cos \sqrt{99}\,t)$$
$$- e^{-t}(c_1 \cos \sqrt{99}\,t + c_2 \sin \sqrt{99}\,t)$$
$$+ 0.0874 \sin 5t + 0.655 \cos 5t$$

Initial conditions: if $t = 0$, $x = 0.25$ and $\frac{dx}{dt} = 0$.

(1) $0.25 = c_1 - 0.01747$,   or $c_1 = 0.267$

(2) $0 = \sqrt{99}\,c_2 - c_1 + 0.655$

  $0 = \sqrt{99}\,c_2 - 0.267 + 0.655$,   or $c_2 = -0.0390$

Thus

  $x_c = e^{-t}(0.267 \cos \sqrt{99}\,t - 0.0390 \sin \sqrt{99}\,t)$

  $x_p = -0.01747 \cos 5t + 0.131 \sin 5t$

Using two significant digits,

  $x_c = e^{-1.0t}(0.27 \cos 9.9t - 0.039 \sin 9.9t)$

  $x_p = -0.017 \cos 5t + 0.13 \sin 5t$

21. $$L \frac{d^2q}{dt^2} + \frac{1}{C}q = e(t)$$

$$0.5 \frac{d^2q}{dt^2} + \frac{1}{8 \times 10^{-4}}q = 50 \sin 100t$$

$$\frac{d^2q}{dt^2} + 2500q = 100 \sin 100t \qquad (*)$$

q_c: $q_c = c_1 \cos 50t + c_2 \sin 50t$

q_p: $q_p = A \cos 100t + B \sin 100t$

$q_p' = -100A \sin 100t + 100B \cos 100t$
$q_p'' = -10000A \cos 100t - 10000B \sin 100t$

Substituting in equation (*):

$(-10000A \cos 100t - 10000 B \sin 100t)$
$$+ 2500(A \cos 100t + B \sin 100t) = 100 \sin 100t$$
$(-10000 + 2500)B \sin 100t + (-10000 + 2500)A \cos 100t$
$$= 100 \sin 100t$$

$-7500B = 100, \quad B = -\frac{1}{75}$

$\quad A = 0$

General solution:

$q = c_1 \cos 50t + c_2 \sin 50t - \left(\frac{1}{75}\right)\sin 100t$

$\frac{dq}{dt} = -50c_1 \sin 50t + 50c_2 \cos 50t - \left(\frac{4}{3}\right)\cos 100t$

Initial conditions: if $t = 0$, then $q = 0$ and $\frac{dq}{dt} = 0$.

(1) $\quad 0 = c_1$

(2) $\quad 0 = 50c_2 - \frac{4}{3}$, or $c_2 = \frac{2}{75}$

Thus

$q(t) = \left(\frac{2}{75}\right)\sin 50t - \left(\frac{1}{75}\right)\sin 100t$

25. $\frac{d^2q}{dt^2} + 10 \frac{dq}{dt} + 100q = 50 \cos 10t$

$q_p = A \cos 10t + B \sin 10t$

$q_p' = -10A \sin 10t + 10B \cos 10t$

$q_p'' = -100A \cos 10t - 100B \sin 10t$

Substituting:

$(-100A \cos 10t - 100B \sin 10t) + 10(-10A \sin 10t + 10B \cos 10t)$
$$+ 100(A \cos 10t + B \sin 10t) = 50 \cos 10t$$

or

$$100B \cos 10t - 100A \sin 10t = 50 \cos 10t$$

$B = \frac{1}{2}, \quad A = 0$

Thus

$q_p = \frac{1}{2} \sin 10t$

and

$i_p = 5 \cos 10t$

1. $D^4 y = 0$, $m^4 = 0$, so that $m = 0, 0, 0, 0$.

$y = c_1 + c_2 x + c_3 x^2 + c_4 x^3$

5. $(D^2 - 2D - 2)y = 0$; $m^2 - 2m - 2 = 0$

$$m = \frac{2 \pm \sqrt{4 + 8}}{2} = \frac{2 \pm 2\sqrt{3}}{2}$$

$$= 1 \pm \sqrt{3}$$

$y = c_1 e^{(1+\sqrt{3})x} + c_2 e^{(1-\sqrt{3})x} = e^x (c_1 e^{\sqrt{3}x} + c_2 e^{-\sqrt{3}x})$

9. $(m - 2)^2 (m^2 + 1) = 0$

$$m = 2, 2, \pm j$$

$y = c_1 e^{2x} + c_2 x e^{2x} + c_3 \cos x + c_4 \sin x$

13. $(3D^2 - 2D - 2)y = 0$

$3m^2 - 2m - 2 = 0$

$$m = \frac{2 \pm \sqrt{4 + 24}}{6} = \frac{2 \pm 2\sqrt{7}}{6} = \frac{1}{3} \pm \frac{1}{3}\sqrt{7}$$

$y = c_1 e^{(1/3+\sqrt{7}/3)x} + c_2 e^{(1/3-\sqrt{7}/3)x}$

$= e^{(1/3)x} \left(c_1 e^{(\sqrt{7}/3)x} + c_2 e^{-(\sqrt{7}/3)x} \right)$

17. $(D^2 - 3D - 4)y = 2 \sin x$

y_c: $m^2 - 3m - 4 = 0$

. $(m - 4)(m + 1) = 0$

$$m = 4, -1$$

$y_c = c_1 e^{4x} + c_2 e^{-x}$

y_p: $y_p = A \cos x + B \sin x$

$y_p' = -A \sin x + B \cos x$

$y_p'' = -A \cos x - B \sin x$

Substituting in the equation:

$(-A \cos x - B \sin x) - 3(-A \sin x + B \cos x)$

$- 4(A \cos x + B \sin x) = 2 \sin x$

$(-5A - 3B)\cos x + (3A - 5B)\sin x = 2 \sin x$

$$-5A - 3B = 0$$

$$3A - 5B = 2$$

Solution: $B = -\frac{5}{17}$, $A = \frac{3}{17}$

Thus

$y = c_1 e^{4x} + c_2 e^{-x} + \frac{3}{17}\cos x - \frac{5}{17}\sin x$

21. $(D^2 - 2D - 3)y = 0$

$$m^2 - 2m - 3 = 0$$
$$(m - 3)(m + 1) = 0$$
$$m = 3, -1$$

$y = c_1 e^{3x} + c_2 e^{-x}$
$Dy = 3c_1 e^{3x} - c_2 e^{-x}$
If $x = 0$, then $y = 0$ and $Dy = -4$:

(1)　　$0 = c_1 + c_2$

(2)　　$\underline{-4 = 3c_1 - c_2}$
　　　　$-4 = 4c_1$, or $c_1 = -1$; $c_2 = 1$

Thus

$y = e^{-x} - e^{3x}$

25. By Hooke's law,

$$F = kx$$
$$5 = k \cdot \frac{1}{2} \quad \text{or} \quad k = 10 \qquad\qquad \text{spring constant}$$

mass:　$\frac{5}{32}$ slug　　　　　　　　　　　　　　　　　　　mass

$$m\,\frac{d^2x}{dt^2} + kx = f(t) \qquad\qquad b = 0$$

$$\frac{5}{32}\,\frac{d^2x}{dt^2} + 10x = \frac{1}{8}\cos 4t \qquad\qquad f(t) = \frac{1}{8}\cos 4t$$

(*)　　　$\dfrac{d^2x}{dt^2} + 64x = \dfrac{4}{5}\cos 4t$ 　　　　　multiplying by $\frac{32}{5}$

x_c:　$m^2 + 64 = 0$, $m = \pm 8j$

$x_c = c_1 \cos 8t + c_2 \sin 8t$

x_p:　$x_p = A\cos 4t + B\sin 4t$

$x_p' = -4A\sin 4t + 4B\cos 4t$

$x_p'' = -16A\cos 4t - 16B\sin 4t$

Substituting in (*):

$(-16A\cos 4t - 16B\sin 4t)$

　　　$+ 64(A\cos 4t + B\sin 4t) = \frac{4}{5}\cos 4t$

250

or

$$48A \cos 4t + 48B \sin 4t = \frac{4}{5} \cos 4t$$

It follows that
$$B = 0 \text{ and } A = \frac{4}{5} \cdot \frac{1}{48} = \frac{1}{60}$$

General solution:
$$x(t) = x_c + x_p = c_1 \cos 8t + c_2 \sin 8t + \frac{1}{60} \cos 4t$$
$$x'(t) = -8c_1 \sin 8t + 8c_2 \cos 8t - \frac{1}{15} \sin 4t$$

Initial conditions:

(1)   if $t = 0$, $x = \frac{1}{4}$ ft   initial position

(2)   if $t = 0$, $\frac{dx}{dt} = -5$ ft/s   initial velocity

Substituting:

$$\frac{1}{4} = c_1 + 0 + \frac{1}{60}, \quad \text{or} \quad c_1 = \frac{1}{4} - \frac{1}{60} = \frac{15}{60} - \frac{1}{60}$$
$$= \frac{14}{60} = \frac{7}{30}$$

$$-5 = 0 + 8c_2 + 0, \text{ or } c_2 = -\frac{5}{8}$$

Substituting the values of c_1 and c_2, we get

$$x(t) = \frac{7}{30} \cos 8t - \frac{5}{8} \sin 8t + \frac{1}{60} \cos 4t$$

CHAPTER 13   THE LAPLACE TRANSFORM

<u>Sections 13.1-13.3</u>

5. $f(t) = t + \cos 2t$

$F(s) = \dfrac{1}{s^2} + \dfrac{s}{s^2 + 4}$ by transforms 2 and 6, respectively.

Thus

$F(s) = \dfrac{s^2 + 4 + s^3}{s^2(s^2 + 4)} = \dfrac{s^3 + s^2 + 4}{s^2(s^2 + 4)}$

9. $f(t) = t^3 e^{-4t}$. By transform 9 with n = 3 and a = -4:

$F(s) = \dfrac{3!}{(s + 4)^4} = \dfrac{6}{(s + 4)^4}$

13. $f(t) = 4 - 5 \sin 2t$. By transforms 1 and 5, respectively:

$F(s) = \dfrac{4}{s} - 5 \dfrac{2}{s^2 + 4} = \dfrac{4}{s} - \dfrac{10}{s^2 + 4}$

17. $F(s) = \dfrac{1}{(s + 2)^3}$. By transform 9 with n = 2 and a = -2:

$F(s) = \dfrac{1}{2} \dfrac{2!}{(s + 2)^3}$ and $f(t) = \dfrac{1}{2} t^2 e^{-2t}$

21. $F(s) = \dfrac{4s + 2}{(s^2 + 4)^2} = \dfrac{4s}{(s^2 + 4)^2} + \dfrac{2}{(s^2 + 4)^2}$

By transform 13 with a = 2:

$\pounds^{-1}\left\{\dfrac{4s}{(s^2 + 4)^2}\right\} = \pounds^{-1}\left\{\dfrac{2(2s)}{(s^2 + 4)^2}\right\} = t \sin 2t$

By transform 12 with a = 2:

$\pounds^{-1}\left\{\dfrac{2}{(s^2 + 4)^2}\right\} = \dfrac{1}{8}\pounds^{-1}\left\{\dfrac{2 \cdot 2^3}{(s^2 + 4)^2}\right\} = \dfrac{1}{8}(\sin 2t - 2t \cos 2t)$

Combining, we get:

$f(t) = t \sin 2t + \dfrac{1}{8} \sin 2t - \dfrac{1}{4} t \cos 2t$

$\qquad = \left(t + \dfrac{1}{8}\right)\sin 2t - \dfrac{1}{4} t \cos 2t$

25. $F(s) = \dfrac{2}{s^2(s^2 + 4)}$. By transform 11 with a = 2:

$F(s) = \dfrac{1}{4}\dfrac{2^3}{s^2(s^2 + 4)}$ and $f(t) = \dfrac{1}{4}(2t - \sin 2t)$

$$= \dfrac{1}{2}t - \dfrac{1}{4}\sin 2t$$

29. By transform 7 with a = -3 and b = $\sqrt{5}$:

$F(s) = \dfrac{1}{\sqrt{5}}\dfrac{\sqrt{5}}{(s + 3)^2 + 5}$ and $f(t) = \dfrac{1}{\sqrt{5}}e^{-3t}\sin\sqrt{5}t$

33. $\mathcal{L}^{-1}\left\{\dfrac{s}{s^2 - 6s + 10}\right\} = \mathcal{L}^{-1}\left\{\dfrac{s}{(s - 3)^2 + 1}\right\} = \mathcal{L}^{-1}\left\{\dfrac{(s - 3 + 3)}{(s - 3)^2 + 1}\right\}$

$$= \mathcal{L}^{-1}\left\{\dfrac{s - 3}{(s - 3)^2 + 1}\right\} + \mathcal{L}^{-1}\left\{\dfrac{3}{(s - 3)^2 + 1}\right\}$$

$$= \mathcal{L}^{-1}\left\{\dfrac{s - 3}{(s - 3)^2 + 1}\right\} + 3\mathcal{L}^{-1}\left\{\dfrac{1}{(s - 3)^2 + 1}\right\}$$

$$= e^{3t}\cos t + 3e^{3t}\sin t = e^{3t}(\cos t + 3\sin t)$$

37. $\dfrac{1}{s(s + 1)}$. Rule I, distinct linear factors:

$\dfrac{A}{s} + \dfrac{B}{s + 1} = \dfrac{A(s + 1) + Bs}{s(s + 1)}$

$A(s + 1) + Bs = 1$

Let s = 0: $A(1) + 0 = 1$

$A = 1$

Let s = -1: $0 + B(-1) = 1$

$B = -1$

$\mathcal{L}^{-1}\left\{\dfrac{1}{s(s + 1)}\right\} = \mathcal{L}^{-1}\left\{\dfrac{1}{s} - \dfrac{1}{s + 1}\right\} = 1 - e^{-t}$

by transforms 1 and 4, respectively.

41. $\dfrac{s^2}{(s - 2)(s + 2)(s - 4)}$. Rule I, distinct linear factors:

$\dfrac{A}{s - 2} + \dfrac{B}{s + 2} + \dfrac{C}{s - 4}$

$= \dfrac{A(s + 2)(s - 4) + B(s - 2)(s - 4) + C(s - 2)(s + 2)}{(s - 2)(s + 2)(s - 4)}$

$A(s + 2)(s - 4) + B(s - 2)(s - 4) + C(s - 2)(s + 2) = s^2$

Let s = - 2: $0 + B(-4)(-6) + 0 = (-2)^2$

$24B = 4$

$B = \dfrac{1}{6}$

Let s = 4: $0 + 0 + C(2)(6) = 4^2$

$$12C = 16$$
$$C = \frac{4}{3}$$

Let s = 2: $A(4)(-2) + 0 + 0 = 2^2$

$$-8A = 4$$
$$A = -\frac{1}{2}$$

$$\mathcal{L}^{-1}\left\{-\frac{1}{2}\,\frac{1}{s-2} + \frac{1}{6}\,\frac{1}{s+2} + \frac{4}{3}\,\frac{1}{s-4}\right\} = -\frac{1}{2}e^{2t} + \frac{1}{6}e^{-2t} + \frac{4}{3}e^{4t}$$

by tranform 4.

45. $\dfrac{1}{(s+1)(s^2+1)}$. (Distinct factors, one linear, one quadratic)

By Rules I and II:

$$\frac{A}{s+1} + \frac{Bs+C}{s^2+1} = \frac{A(s^2+1) + (Bs+C)(s+1)}{(s+1)(s^2+1)}$$

$$A(s^2+1) + (Bs+C)(s+1) = 1$$

Let s = -1: $A(2) + 0 = 1$

$$A = \frac{1}{2}$$

Let s = 0: $\frac{1}{2}(1) + C(1) = 1$ (since $A = \frac{1}{2}$)

$$C = \frac{1}{2}$$

Let s = 1: $\frac{1}{2}(2) + \left(B + \frac{1}{2}\right)(2) = 1$ $A = \frac{1}{2}, \; C = \frac{1}{2}$

$$1 + \left(B + \frac{1}{2}\right)(2) = 1$$

$$\left(B + \frac{1}{2}\right)(2) = 0$$

$$B = -\frac{1}{2}$$

$$\mathcal{L}^{-1}\left\{\frac{1}{2}\,\frac{1}{s+1} - \frac{1}{2}\,\frac{s}{s^2+1} + \frac{1}{2}\,\frac{1}{s^2+1}\right\} = \frac{1}{2}e^{-t} - \frac{1}{2}\cos t + \frac{1}{2}\sin t$$

49. $\dfrac{1}{(s^2+1)(s-1)^2}$. By Rule I, repeating linear factors, and Rule II:

$$\frac{As+B}{s^2+1} + \frac{C}{s-1} + \frac{D}{(s-1)^2}$$

$$= \frac{(As+B)(s-1)^2 + C(s^2+1)(s-1) + D(s^2+1)}{(s^2+1)(s-1)^2}$$

$$(As+B)(s-1)^2 + C(s^2+1)(s-1) + D(s^2+1) = 1$$

Let s = 1: $0 + 0 + D(2) = 1$

$$D = \frac{1}{2}$$

$$(As+B)(s-1)^2 + C(s^2+1)(s-1) + \frac{1}{2}(s^2+1) = 1$$

Let s = 0, 2, and -1, respectively:

$$B - C + \frac{1}{2} = 1$$
$$2A + B + 5C + \frac{5}{2} = 1$$
$$\underline{-4A + 4B - 4C + 1 = 1}$$

(1) $\qquad B - C \qquad = \frac{1}{2}$

(2) $\quad 2A + B + 5C \qquad = -\frac{3}{2}$

(3) $\quad -4A + 4B - 4C \qquad = 0$

(4) $\qquad B - C \qquad = \frac{1}{2}$

(5) $\quad 4A + 2B + 10C \qquad = -3 \qquad$ multiplying (2) by 2

(6) $\quad -4A + 4B - 4C \qquad = 0$

(7) $\qquad B - C \qquad = \frac{1}{2}$

(8) $\qquad 6B + 6C \qquad = -3 \qquad$ adding (5) and (6)

$\qquad\qquad 6B - 6C \qquad = 3 \qquad$ multiplying (7) by (6)

$\qquad\qquad \underline{6B + 6C \qquad = -3}$

$\qquad\qquad 12B \qquad\qquad = 0 \qquad$ adding

$$B = 0$$
$$C = -\frac{1}{2} \qquad \text{by (1)}$$

By (2) above, we get A = $\frac{1}{2}$.

$$\mathcal{L}^{-1}\left\{\frac{1}{2}\frac{s}{s^2 + 1} - \frac{1}{2}\frac{1}{s - 1} + \frac{1}{2}\frac{1}{(s - 1)^2}\right\}$$

$$= \frac{1}{2}\cos t - \frac{1}{2}e^t + \frac{1}{2}te^t$$

53.
$$\frac{2s^2 + 2s + 1}{(s^2 + 2s + 2)(s - 1)} = \frac{As + B}{s^2 + 2s + 2} + \frac{C}{s - 1}$$

$$= \frac{(As + B)(s - 1) + C(s^2 + 2s + 2)}{(s^2 + 2s + 2)(s - 1)}$$

$$(As + B)(s - 1) + C(s^2 + 2s + 2) = 2s^2 + 2s + 1$$

Let s = 1: $\quad 0 + C(5) = 5$

$$\qquad\qquad\qquad C = 1$$

Let s = 0: $\quad B(-1) + 1(2) = 1 \qquad$ (since C = 1)

$$\qquad\qquad\qquad B = 1$$

Let s = 2: $\quad (2A + 1)(1) + 1(10) = 13 \qquad$ B = 1, C = 1

$$\qquad\qquad 2A + 1 = 3$$

$$\qquad\qquad A = 1$$

$$\mathcal{L}^{-1}\left\{\frac{s + 1}{(s + 1)^2 + 1} + \frac{1}{s - 1}\right\} = e^{-t}\cos t + e^t$$

255

1. $y' - y = 0$, $y(0) = 1$

    $sY(s) - y(0) - Y(s) = 0$ by (13.11)

      $sY(s) - 1 - Y(s) = 0$ since $y(0) = 1$

        $sY(s) - Y(s) = 1$

          $(s - 1)Y(s) = 1$ factoring $Y(s)$

              $Y(s) = \dfrac{1}{s - 1}$ dividing by $s - 1$

                $y = e^t$ by transform 4

5. $y' - 2y = e^{2t}$, $y(0) = 0$

    $sY(s) - y(0) - 2Y(s) = \dfrac{1}{s - 2}$

        $sY(s) - 2Y(s) = \dfrac{1}{s - 2}$ $y(0) = 0$

         $(s - 2)Y(s) = \dfrac{1}{s - 2}$ factoring $Y(s)$

             $Y(s) = \dfrac{1}{(s - 2)^2}$ dividing by $s - 2$

                $y = te^{2t}$ by transform 9

9. $y'' + 9y = 0$, $y(0) = 1$, $y'(0) = -2$

    $s^2Y(s) - sy(0) - y'(0) + 9Y(s) = 0$ by (13.12)

        $s^2Y(s) - s + 2 + 9Y(s) = 0$ $y(0) = 1$, $y'(0) = -2$

           $s^2Y(s) + 9Y(s) = s - 2$

             $(s^2 + 9)Y(s) = s - 2$

                 $Y(s) = \dfrac{s - 2}{s^2 + 9} = \dfrac{s}{s^2 + 9} - \dfrac{2}{s^2 + 9}$

                      $= \dfrac{s}{s^2 + 9} - 2\dfrac{1}{s^2 + 9}$

                      $= \dfrac{s}{s^2 + 9} - \dfrac{2}{3}\dfrac{3}{s^2 + 9}$

                  $y = \cos 3t - \dfrac{2}{3} \sin 3t$

13. $y' - y = \cos 2t$, $y(0) = 0$

    $sY(s) - y(0) - Y(s) = \dfrac{s}{s^2 + 4}$ by transform 6

        $sY(s) - Y(s) = \dfrac{s}{s^2 + 4}$ $y(0) = 0$

         $(s - 1)Y(s) = \dfrac{s}{s^2 + 4}$

             $Y(s) = \dfrac{s}{(s^2 + 4)(s - 1)}$

$Y(s) = \dfrac{As + B}{s^2 + 4} + \dfrac{C}{s - 1} = \dfrac{(As + B)(s - 1) + C(s^2 + 4)}{(s^2 + 4)(s - 1)}$

$$(As + B)(s - 1) + C(s^2 + 4) = s$$

Let $s = 1$: $\quad 0 + C(5) = 1$
$$C = \tfrac{1}{5}$$

Let $s = 0$: $\quad B(-1) + \tfrac{1}{5}(4) = 0 \qquad\qquad$ since $C = \tfrac{1}{5}$
$$B = \tfrac{4}{5}$$

Let $s = 2$: $\quad \left(2A + \tfrac{4}{5}\right)(1) + \tfrac{1}{5}(8) = 2 \qquad\qquad C = \tfrac{1}{5}, \ B = \tfrac{4}{5}$
$$A = -\tfrac{1}{5}$$

$$Y(s) = -\tfrac{1}{5}\,\frac{s}{s^2 + 4} + \tfrac{4}{5}\,\frac{1}{s^2 + 4} + \tfrac{1}{5}\,\frac{1}{s - 1}$$

$$= -\tfrac{1}{5}\,\frac{s}{s^2 + 4} + \tfrac{4}{5}\,\tfrac{1}{2}\,\frac{2}{s^2 + 4} + \tfrac{1}{5}\,\frac{1}{s - 1}$$

$$y = -\tfrac{1}{5}\cos 2t + \tfrac{2}{5}\sin 2t + \tfrac{1}{5}e^t$$

17. $\quad y'' + 6y' + 13y = 0, \ y(0) = 1, \ y'(0) = -2$

$$s^2 Y(s) - sy(0) - y'(0) + 6(sY(s) - y(0)) + 13Y(s) = 0$$
$$s^2 Y(s) - s + 2 + 6(sY(s) - 1) + 13Y(s) = 0$$
$$s^2 Y(s) + 6sY(s) + 13Y(s) = s + 4$$
$$(s^2 + 6s + 13)Y(s) = s + 4$$

$$Y(s) = \frac{s + 4}{s^2 + 6s + 13} = \frac{s + 4}{(s + 3)^2 + 4} = \frac{(s + 3) + 1}{(s + 3)^2 + 4}$$

$$= \frac{s + 3}{(s + 3)^2 + 4} + \frac{1}{(s + 3)^2 + 4}$$

$$= \frac{s + 3}{(s + 3)^2 + 4} + \tfrac{1}{2}\,\frac{2}{(s + 3)^2 + 4}$$

$$y = e^{-3t}\cos 2t + \tfrac{1}{2}e^{-3t}\sin 2t$$

21. $\quad y'' - 4y = 3\cos t, \ y(0) = y'(0) = 0$

$$s^2 Y(s) - sy(0) - y'(0) - 4Y(s) = \frac{3s}{s^2 + 1}$$

$$(s^2 - 4)Y(s) = \frac{3s}{s^2 + 1}$$

$$Y(s) = \frac{3s}{(s^2 - 4)(s^2 + 1)}$$

$$Y(s) = \frac{3s}{(s - 2)(s + 2)(s^2 + 1)}$$

$$\frac{3s}{(s - 2)(s + 2)(s^2 + 1)} = \frac{A}{s - 2} + \frac{B}{s + 2} + \frac{Cs + D}{s^2 + 1}$$

$$= \frac{A(s + 2)(s^2 + 1) + B(s - 2)(s^2 + 1) + (Cs + D)(s - 2)(s + 2)}{(s - 2)(s + 2)(s^2 + 1)}$$

$$A(s + 2)(s^2 + 1) + B(s - 2)(s^2 + 1) + (Cs + D)(s - 2)(s + 2) = 3s$$

Let $s = 2$: $\quad A(4)(5) + 0 + 0 = 6$
$$A = \frac{3}{10}$$

Let $s = -2$: $\quad 0 + B(-4)(5) + 0 = -6$
$$B = \frac{3}{10}$$

Let $s = 0$: $\quad \frac{3}{10}(2)(1) + \frac{3}{10}(-2)(1) + D(-2)(2) = 0 \qquad A = B = \frac{3}{10}$
$$D = 0$$

Let $s = 1$: $\quad \frac{3}{10}(3)(2) + \frac{3}{10}(-1)(2) + C(-1)(3) = 3$
$$\frac{9}{5} - \frac{3}{5} - 3C = 3$$
$$-3C = \frac{15}{5} - \frac{9}{5} + \frac{3}{5} = \frac{9}{5}$$
$$C = -\frac{3}{5}$$

$$Y(s) = \frac{3}{10}\frac{1}{s - 2} + \frac{3}{10}\frac{1}{s + 2} - \frac{3}{5}\frac{s}{s^2 + 1}$$

$$y = \frac{3}{10}e^{2t} + \frac{3}{10}e^{-2t} - \frac{3}{5}\cos t$$

25. $\quad L\dfrac{d^2q}{dt^2} + R\dfrac{dq}{dt} + \dfrac{1}{C}q = e(t)$

$0.1\dfrac{d^2q}{dt^2} + 6.0\dfrac{dq}{dt} + \dfrac{1}{0.02}q = 6.0$

Omitting final zeros, we get

$0.1\dfrac{d^2q}{dt^2} + 6\dfrac{dq}{dt} + 50q = 6$

$\dfrac{d^2q}{dt^2} + 60\dfrac{dq}{dt} + 500q = 60$

$s^2Q(s) - sq(0) - q'(0) + 60(sQ(s) - q(0)) + 500Q(s) = \dfrac{60}{s}$

$$s^2Q(s) + 60sQ(s) + 500Q(s) = \frac{60}{s}$$

$Q(s) = \dfrac{60}{s(s^2 + 60s + 500)} = \dfrac{60}{s(s + 50)(s + 10)}$

$\quad = \dfrac{A}{s} + \dfrac{B}{s + 50} + \dfrac{C}{s + 10} = \dfrac{3}{25}\dfrac{1}{s} + \dfrac{3}{100}\dfrac{1}{s + 50} - \dfrac{3}{20}\dfrac{1}{s + 10}$

$q(t) = \dfrac{3}{25} + \dfrac{3}{100}e^{-50t} - \dfrac{3}{20}e^{-10t}$

$\quad = 0.12 + 0.03e^{-50t} - 0.15e^{-10t}$

1. $f(t) = 2e^{-3t}$

   $F(s) = 2 \dfrac{1}{s+3} = \dfrac{2}{s+3}$ by transform 4

5. $f(t) = 2t - \sin 2t$

   $F(s) = \dfrac{2}{s^2} - \dfrac{2}{s^2+4}$ by transforms 2 and 5, respectively.

   Thus

   $F(s) = \dfrac{2(s^2+4) - 2s^2}{s^2(s^2+4)} = \dfrac{8}{s^2(s^2+4)}$

9. $F(s) = \dfrac{s}{s^2 - 2s + 5} = \dfrac{s}{(s-1)^2 + 4} = \dfrac{(s-1)+1}{(s-1)^2 + 4}$

   $= \dfrac{s-1}{(s-1)^2 + 4} + \dfrac{1}{2}\dfrac{2}{(s-1)^2 + 4}$

   $f(t) = e^t \cos 2t + \dfrac{1}{2}e^t \sin 2t$

   $= e^t(\cos 2t + \dfrac{1}{2}\sin 2t)$

13. $F(s) = \dfrac{1}{(s+2)(s-3)(s-4)} = \dfrac{A}{s+2} + \dfrac{B}{s-3} + \dfrac{C}{s-4}$

    $= \dfrac{A(s-3)(s-4) + B(s+2)(s-4) + C(s+2)(s-3)}{(s+2)(s-3)(s-4)}$

    $A(s-3)(s-4) + B(s+2)(s-4) + C(s+2)(s-3) = 1$

    Let $s = 3$: $B(5)(-1) = 1$

    $\qquad\qquad\qquad B = -\dfrac{1}{5}$

    Let $s = 4$: $C(6)(1) = 1$

    $\qquad\qquad\qquad C = \dfrac{1}{6}$

    Let $s = -2$: $A(-5)(-6) = 1$

    $\qquad\qquad\qquad A = \dfrac{1}{30}$

    $F(s) = \dfrac{1}{30}\dfrac{1}{s+2} - \dfrac{1}{5}\dfrac{1}{s-3} + \dfrac{1}{6}\dfrac{1}{s-4}$

    $f(t) = \dfrac{1}{30}e^{-2t} - \dfrac{1}{5}e^{3t} + \dfrac{1}{6}e^{4t}$

17. $y' + 2y = te^{-2t}$, $y(0) = -1$

    $sY(s) - y(0) + 2Y(s) = \dfrac{1}{(s+2)^2}$ by transform 9

    $sY(s) + 1 + 2Y(s) = \dfrac{1}{(s+2)^2}$ $y(0) = -1$

259

$$(s + 2)Y(s) = \frac{1}{(s + 2)^2} - 1$$

$$Y(s) = \frac{1}{(s + 2)^3} - \frac{1}{s + 2} = \frac{1}{2} \frac{2!}{(s + 2)^3} - \frac{1}{s + 2}$$

$$y = \frac{1}{2}t^2 e^{-2t} - e^{-2t} = e^{-2t}\left(\frac{1}{2}t^2 - 1\right)$$

21. $y'' + 2y' + 5y = 0$, $y(0) = 1$, $y'(0) = 0$

$$s^2Y(s) - sy(0) - y'(0) + 2(sY(s) - y(0)) + 5Y(s) = 0$$

$$s^2Y(s) - s + 2(sY(s) - 1) + 5Y(s) = 0$$

$$s^2Y(s) - s + 2sY(s) - 2 + 5Y(s) = 0$$

$$(s^2 + 2s + 5)Y(s) = s + 2$$

$$Y(s) = \frac{s + 2}{s^2 + 2s + 5}$$

$$Y(s) = \frac{s + 2}{(s + 1)^2 + 4} = \frac{(s + 1) + 1}{(s + 1)^2 + 4}$$

$$= \frac{s + 1}{(s + 1)^2 + 4} + \frac{1}{2} \frac{2}{(s + 1)^2 + 4}$$

$$y = e^{-t} \cos 2t + \frac{1}{2}e^{-t} \sin 2t$$

$$= e^{-t}\left(\cos 2t + \frac{1}{2} \sin 2t\right)$$

25. By Hooke's law,

$$F = kx$$
$$4 = k \cdot \frac{1}{2} \quad \text{or} \quad k = 8$$

mass: $\frac{4}{32} = \frac{1}{8}$ slug

equation:

$$\frac{1}{8} \frac{d^2x}{dt^2} + 2 \frac{dx}{dt} + 8x = 0 \qquad\qquad \text{since } b = 2$$

and

$$\frac{d^2x}{dt^2} + 16 \frac{dx}{dt} + 64x = 0$$

Initial conditions:

(1) if $t = 0$, $x = 0$ initial position

(2) if $t = 0$, $\frac{dx}{dt} = -4$ initial velocity

$$s^2X(s) - sx(0) - x'(0) + 16(sX(s) - x(0)) + 64X(s) = 0$$

$$s^2X(s) - (-4) + 16sX(s) + 64X(s) = 0$$

$$X(s)(s^2 + 16s + 64) = -4$$

$$X(s) = \frac{-4}{s^2 + 16s + 64} = -\frac{4}{(s + 8)^2}$$

$$x(t) = -4te^{-8t} \text{ by transform 9}$$